E
SOURCE
The Prentice Hall
ENGINEERING SOURCE

Engineering Ethics

Charles B. Fleddermann

Department of Electrical and Computer Engineering
University of New Mexico

Prentice Hall
Upper Saddle River, NJ 07458

Library of Congress Information Available

Editor-in-chief: **MARCIA HORTON**
Acquisitions editor: **ERIC SVENDSEN**
Director of production and manufacturing: **DAVID W. RICCARDI**
Managing editor: **EILEEN CLARK**
Editorial/production supervision: **ROSE KERNAN**
Cover director: **JAYNE CONTE**
Creative director: **AMY ROSEN**
Marketing manager: **DANNY HOYT**
Manufacturing buyer: **PAT BROWN**
Editorial assistant: **GRIFFIN CABLE**

Reprinted with corrections July, 1999

Printed in the United States of America

10 9 8 7 6 5 4 3 2 1

ISBN 0-13-784224-4

Prentice-Hall International (UK) Limited, *London*
Prentice-Hall of Australia Pty. Limited, *Sydney*
Prentice-Hall Canada, Inc., *Toronto*
Prentice-Hall Hispanoamericana, S.A., *Mexico*
Prentice-Hall of India Private Limited, *New Delhi*
Prentice-Hall of Japan, Inc., *Tokyo*
Prentice-Hall Asia Pte., Ltd., *Singapore*
Editora Prentice-Hall do Brazil, Ltda., *Rio de Janeiro*

About ESource

The Challenge

Professors who teach the Introductory/First-Year Engineering course popular at most engineering schools have a unique challenge—teaching a course defined by a changing curriculum. The first-year engineering course is different from any other engineering course in that there is no real cannon that defines the course content. It is not like Engineering Mechanics or Circuit Theory where a consistent set of topics define the course. Instead, the introductory engineering course is most often defined by the creativity of professors and students, and the specific needs of a college or university each semester. Faculty involved in this course typically put extra effort into it, and it shows in the uniqueness of each course at each school.

Choosing a textbook can be a challenge for unique courses. Most freshmen require some sort of reference material to help them through their first semesters as a college student. But because faculty put such a strong mark on their course, they often have a difficult time finding the right mix of materials for their course and often have to go without a text, or with one that does not really fit. Conventional textbooks are far too static for the typical specialization of the first-year course. How do you find the perfect text for your course that will support your students educational needs, but give you the flexibility to maximize the potential of your course?

ESource—The Prentice Hall Engineering Source http://emissary.prenhall.com/esource

Prentice Hall created ESource—The Prentice-Hall Engineering Source—to give professors the power to harness the full potential of their text and their freshman/first year engineering course. In today's technologically advanced world, why settle for a book that isn't perfect for your course? Why not have a book that has the exact blend of topics that you want to cover with your students?

More then just a collection of books, ESource is a unique publishing system revolving around the ESource website—http://emissary.prenhall.com/esource/. ESource enables you to put your stamp on your book just as you do your course. It lets you:

Control You choose exactly what chapters or sections are in your book and in what order they appear. Of course, you can choose the entire book if you'd like and stay with the author's original order.

Optimize Get the most from your book and your course. ESource lets you produce the optimal text for your students needs.

Customize You can add your own material anywhere in your text's presentation, and your final product will arrive at your bookstore as a professionally formatted text.

ESource Content

All the content in ESource was written by educators specifically for freshman/first-year students. Authors tried to strike a balanced level of presentation, one that was not either too formulaic and trivial, but not focusing heavily on advanced topics that most introductory students will not encounter until later classes. A developmental editor reviewed the books and made sure that every text was written at the appropriate level, and that the books featured a balanced presentation. Because many professors do not have extensive time to cover these topics in the classroom, authors prepared each text with the idea that many students would use it for self-instruction and independent study. Students should be able to use this content to learn the software tool or subject on their own.

While authors had the freedom to write texts in a style appropriate to their particular subject, all followed certain guidelines created to promote the consistency a text needs. Namely, every chapter opens with a clear set of objectives to lead students into the chapter. Each chapter also contains practice problems that tests a student's skill at performing the tasks they have just learned. Chapters close with extra practice questions and a list of key terms for reference. Authors tried to focus on motivating applications that demonstrate how engineers work in the real world, and included these applications throughout the text in various chapter openers, examples, and problem material. Specific Engineering and Science **Application Boxes** are also located throughout the texts, and focus on a specific application and demonstrating its solution.

Because students often have an adjustment from high school to college, each book contains several **Professional Success Boxes** specifically designed to provide advice on college study skills. Each author has worked to provide students with tips and techniques that help a student better understand the material, and avoid common pitfalls or problems first-year students often have. In addition, this series contains an entire book titled ***Engineering Success*** by Peter Schiavone of the University of Alberta intended to expose students quickly to what it takes to be an engineering student.

Creating Your Book

Using ESource is simple. You preview the content either on-line or through examination copies of the books you can request on-line, from your PH sales rep, or by calling(1-800-526-0485). Create an on-line outline of the content you want in the order you want using ESource's simple interface. Either type or cut and paste your own material and insert it into the text flow. You can preview the overall organization of the text you've created at anytime (please note, since this preview is immediate, it comes unformatted.), then press another button and receive an order number for your own custom book . If you are not ready to order, do nothing—ESource will save your work. You can come back at any time and change, re-arrange, or add more material to your creation. You are in control. Once you're finished and you have an ISBN, give it to your bookstore and your book will arrive on their shelves six weeks after the order. Your custom desk copies with their instructor supplements will arrive at your address at the same time.

To learn more about this new system for creating the perfect textbook, go to **http://emissary.prenhall.com/esource/**. You can either go through the on-line walkthrough of how to create a book, or experiment yourself.

http://emissary.prenhall.com/esource/

Community

ESource has two other areas designed to promote the exchange of information among the introductory engineering community, the Faculty and the Student Centers. Created and maintained with the help of Dale Calkins, an Associate Professor at the University of Washington, these areas contain a wealth of useful information and tools. You can preview outlines created by other schools and can see how others organize their courses. Read a monthly article discussing important topics in the curriculum. You can post your own material and share it with others, as well as use what others have posted in your own documents. Communicate with our authors about their books and make suggestions for improvement. Comment about your course and ask for information from others professors. Create an on-line syllabus using our custom syllabus builder. Browse Prentice Hall's catalog and order titles from your sales rep. Tell us new features that we need to add to the site to make it more useful.

Supplements

Adopters of ESource receive an instructor's CD that includes solutions as well as professor and student code for all the books in the series. This CD also contains approximately **350 Powerpoint Transparencies** created by Jack Leifer—of University South Carolina—Aiken. Professors can either follow these transparencies as pre-prepared lectures or use them as the basis for their own custom presentations. In addition, look to the web site to find materials from other schools that you can download and use in your own course.

Titles in the ESource Series

Introduction to Unix
0-13-095135-8
David I. Schwartz

Introduction to Maple
0-13-095133-1
David I. Schwartz

Introduction to Word
0-13-254764-3
David C. Kuncicky

Introduction to Excel
0-13-254749-X
David C. Kuncicky

Introduction to MathCAD
0-13-937493-0
Ronald W. Larsen

Introduction to AutoCAD, R. 14
0-13-011001-9
Mark Dix and Paul Riley

Introduction to the Internet, 2/e
0-13-011037-X
Scott D. James

Design Concepts for Engineers
0-13-081369-9
Mark N. Horenstein

Engineering Design—A Day in the Life of Four Engineers
0-13-085089-6
Mark N. Horenstein

Engineering Ethics
0-13-784224-4
Charles B. Fleddermann

Engineering Success
0-13-080859-8
Peter Schiavone

Mathematics Review
0-13-011501-0
Peter Schiavone

Introduction to ANSI C
0-13-011854-0
Delores M. Etter

Introduction to C++
0-13-011855-9
Delores M. Etter

Introduction to MATLAB
0-13-013149-0
Delores M. Etter

Introduction to FORTRAN 90
0-13-013146-6
Larry Nyhoff and Sanford Leestma

http://emissary.prenhall.com/esource/

About the Authors

No project could ever come to pass without a group of authors who have the vision and the courage to turn a stack of blank paper into a book. The authors in this series worked diligently to produce their books, provide the building blocks of the series.

Delores M. Etter is a Professor of Electrical and Computer Engineering at the University of Colorado. Dr. Etter was a faculty member at the University of New Mexico and also a Visiting Professor at Stanford University. Dr. Etter was responsible for the Freshman Engineering Program at the University of New Mexico and is active in the Integrated Teaching Laboratory at the University of Colorado. She was elected a Fellow of the Institute of Electrical and Electronic Engineers for her contributions to education and for her technical leadership in digital signal processing. IN addition to writing best-selling textbooks for engineering computing, Dr. Etter has also published research in the area of adaptive signal processing.

Sanford Leestma is a Professor of Mathematics and Computer Science at Calvin College, and received his Ph.D from New Mexico State University. He has been the long time co-author of successful textbooks on Fortran, Pascal, and data structures in Pascal. His current research interests are in the areas of algorithms and numerical computation.

Larry Nyhoff is a Professor of Mathematics and Computer Science at Calvin College. After doing bachelors work at Calvin, and Masters work at Michigan, he received a Ph.D. from Michigan State and also did graduate work in computer science at Western Michigan. Dr. Nyhoff has taught at Calvin for the past 34 years—mathematics at first and computer science for the past several years. He has co-authored several computer science textbooks since 1981 including titles on Fortran and C++, as well as a brand new title on Data Structures in C++.

Acknowledgments: We express our sincere appreciation to all who helped in the preparation of this module, especially our acquisitions editor Alan Apt, managing editor Laura Steele, development editor Sandra Chavez, and production editor Judy Winthrop. We also thank Larry Genalo for several examples and exercises and Erin Fulp for the Internet address application in Chapter 10. We appreciate the insightful review provided by Bart Childs. We thank our families—Shar, Jeff, Dawn, Rebecca, Megan, Sara, Greg, Julie, Joshua, Derek, Tom, Joan; Marge, Michelle, Sandy, Lori, Michael—for being patient and understanding. We thank God for allowing us to write this text.

Mark Dix began working with AutoCAD in 1985 as a programmer for CAD Support Associates, Inc. He helped design a system for creating estimates and bills of material directly from AutoCAD drawing databases for use in the automated conveyor industry. This system became the basis for systems still widely in use today. In 1986 he began collaborating with Paul Riley to create AutoCAD training materials, combining Riley's background in industrial design and training with Dix' s background in writing, curriculum development, and programming. Dix and Riley have created tutorial and teaching methods for every AutoCAD release since Version 2.5. Mr. Dix has a Master of Arts in Teaching from Cornell University and a Masters of Education from the University of Massachusetts. He is currently the Director of Dearborn Academy High School in Arlington, Massachusetts.

Paul Riley is an author, instructor, and designer specializing in graphics and design for multimedia. He is a founding partner of CAD Support Associates, a contract service and professional training organization for computer-aided design. His 15 years of business experience and 20 years of teaching experience are supported by degrees

in education and computer science. Paul has taught AutoCAD at the University of Massachusetts at Lowell and is presently teaching AutoCAD at Mt. Ida College in Newton, Massachusetts. He has developed a program, Computer-Aided Design for Professionals that is highly regarded by corporate clients and has been an ongoing success since 1982.

David I. Schwartz is a Lecturer at SUNY-Buffalo who teaches freshman and first-year engineering, and has a Ph.D from SUNY-Buffalo in Civil Engineering. Schwartz originally became interested in Civil engineering out of an interest in building grand structures, but has also pursued other academic interests including artificial intelligence and applied mathematics. He became interested in Unix and Maple through their application to his research, and eventually jumped at the chance to teach these subjects to students. He tries to teach his students to become incremental learners and encourages frequent practice to master a subject, and gain the maturity and confidence to tackle other subjects independently. In his spare time, Schwartz is an avid musician and plays drums in a variety of bands.

Acknowledgments: I would like to thank the entire School of Engineering and Applied Science at the State University of New York at Buffalo for the opportunity to teach not only my students, but myself as well; all my EAS140 students, without whom this book would not be possible—thanks for slugging through my lab packets; Andrea Au, Eric Svendsen, and Elizabeth Wood at Prentice Hall for advising and encouraging me as well as wading through my blizzard of e-mail; Linda and Tony for starting the whole thing in the first place; Rogil Camama, Linda Chattin, Stuart Chen, Jeffrey Chottiner, Roger Christian, Anthony Dalessio, Eugene DeMaitre, Dawn Halvorsen, Thomas Hill, Michael Lamanna, Nate "X" Patwardhan, Durvejai Sheobaran, "Able" Alan Somlo, Ben Stein, Craig Sutton, Barbara Umiker, and Chester "JC" Zeshonski for making this book a reality; Ewa Arrasjid, "Corky" Brunskill, Bob Meyer, and Dave Yearke at "the Department Formerly Known as ECS" for all their friendship, advice, and respect; Jeff, Tony, Forrest, and Mike for the interviews; and, Michael Ryan and Warren Thomas for believing in me.

Ronald W. Larsen is an Associate Professor in Chemical Engineering at Montana State University, and received his Ph.D from the Pennsylvania State University. Larsen was initially attracted to engineering because he felt it was a serving profession, and because engineers are often called on to eliminate dull and routine tasks. He also enjoys the fact that engineering rewards creativity and presents constant challenges. Larsen feels that teaching large sections of students is one of the most challenging tasks he has ever encountered because it enhances the importance of effective communication. He has drawn on a two year experince teaching courses in Mongolia through an interpreter to improve his skills in the classroom. Larsen sees software as one of the changes that has the potential to radically alter the way engineers work, and his book Introduction to Mathcad was written to help young engineers prepare to be productive in an ever-changing workplace.

Acknowledgments: To my students at Montana State University who have endured the rough drafts and typos, and who still allow me to experiment with their classes—my sincere thanks.

Peter Schiavone is a professor and student advisor in the Department of Mechanical Engineering at the University of Alberta. He received his Ph.D. from the University of Strathclyde, U.K. in 1988. He has authored several books in the area of study skills and academic success as well as numerous papers in scientific research journals.

Before starting his career in academia, Dr. Schiavone worked in the private sector for Smith's Industries (Aerospace and Defence Systems Company) and Marconi Instruments in several different areas of engineering including aerospace, systems and software engineering. During that time he developed an interest

http://emissary.prenhall.com/esource/

in engineering research and the applications of mathematics and the physical sciences to solving real-world engineering problems.

His love for teaching brought him to the academic world. He founded the first Mathematics Resource Center at the University of Alberta: a unit designed specifically to teach high school students the necessary survival skills in mathematics and the physical sciences required for first-year engineering. This led to the Students' Union Gold Key award for outstanding contributions to the University and to the community at large.

Dr. Schiavone lectures regularly to freshman engineering students, high school teachers, and new professors on all aspects of engineering success, in particular, maximizing students' academic performance. He wrote the book *Engineering Success* in order to share with you the *secrets of success in engineering study*: the most effective, tried and tested methods used by the most successful engineering students.

Acknowledgments: I'd like to acknowledge the contributions of: Eric Svendsen, for his encouragement and support; Richard Felder for being such an inspiration; the many students who shared their experiences of first-year engineering—both good and bad; and finally, my wife Linda for her continued support and for giving me Conan.

Scott D. James is a staff lecturer at Kettering University (formerly GMI Engineering & Management Institute) in Flint, Michigan. He is currently pursuing a Ph.D. in Systems Engineering with an emphasis on software engineering and computer-integrated manufacturing. Scott decided on writing textbooks after he found a void in the books that were available. "I really wanted a book that showed how to do things in good detail but in a clear and concise way. Many of the books on the market are full of fluff and force you to dig out the really important facts." Scott decided on teaching as a profession after several years in the computer industry. "I thought that it was really important to know what it was like outside of academia. I wanted to provide students with classes that were up to date and provide the information that is really used and needed."

Acknowledgments: Scott would like to acknowledge his family for the time to work on the text and his students and peers at Kettering who offered helpful critique of the materials that eventually became the book.

David C. Kuncicky is a native Floridian. He earned his Baccalaureate in psychology, Master's in computer science, and Ph.D. in computer science from Florida State University. Dr. Kuncicky is the Director of Computing and Multimedia Services for the FAMU-FSU College of Engineering. He also serves as a faculty member in the Department of Electrical Engineering. He has taught computer science and computer engineering courses for the past 15 years. He has published research in the areas of intelligent hybrid systems and neural networks. He is actively involved in the education of computer and network system administrators and is a leader in the area of technology-based curriculum delivery.

Acknowledgments: Thanks to Steffie and Helen for putting up with my late nights and long weekends at the computer. Thanks also to the helpful and insightful technical reviews by the following people: Jerry Ralya, Kathy Kitto of Western Washington University, Avi Singhal of Arizona State University, and Thomas Hill of the State University of New York at Buffalo. I appreciate the patience of Eric Svendsen and Rose Kernan of Prentice Hall for gently guiding me through this project. Finally, thanks to Dean C.J. Chen for providing continued tutelage and support.

Mark Horenstein is an Associate Professor in the Electrical and Computer Engineering Department at Boston University. He received his Bachelors in Electrical Engineering in 1973 from Massachusetts Institute of Technology, his Masters in Electrical Engineering in 1975

from University of California at Berkeley, and his Ph.D. in Electrical Engineering in 1978 from Massachusetts Institute of Technology. Professor Horenstein's research interests are in applied electrostatics and electromagnetics as well as microelectronics, including sensors, instrumentation, and measurement. His research deals with the simulation, test, and measurement of electromagnetic fields. Some topics include electrostatics in manufacturing processes, electrostatic instrumentation, EOS/ESD control, and electromagnetic wave propagation.

Professor Horenstein designed and developed a class at Boston University, which he now teaches entitled Senior Design Project (ENG SC 466). In this course, the student gets real engineering design experience by working for a virtual company, created by Professor Horenstein, that does real projects for outside companies—almost like an apprenticeship. Once in "the company" (Xebec Technologies), the student is assigned to an engineering team of 3-4 persons. A series of potential customers are recruited, from which the team must accept an engineering project. The team must develop a working prototype deliverable engineering system that serves the need of the customer. More than one team may be assigned to the same project, in which case there is competition for the customer's business.

Acknowledgements: Several individuals contributed to the ideas and concepts presented in Design Principles for Engineers. The concept of the Peak Performance design competition, which forms a cornerstone of the book, originated with Professor James Bethune of Boston University. Professor Bethune has been instrumental in conceiving of and running Peak Performance each year and has been the inspiration behind many of the design concepts associated with it. He also provided helpful information on dimensions and tolerance. Several of the ideas presented in the book, particularly the topics on brainstorming and teamwork, were gleaned from a workshop on engineering design help bi-annually by Professor Charles Lovas of Southern Methodist University. The principles of estimation were derived in part from a freshman engineering problem posed by Professor Thomas Kincaid of Boston University.

I would like to thank my family, Roxanne, Rachel, and Arielle, for giving me the time and space to think about and write this book. I also appreciate Roxanne's inspiration and help in identifying examples of human/machine interfaces.

Dedicated to Roxanne, Rachel, and Arielle

Charles B. Fleddermann is a professor in the Department of Electrical and Computer Engineering at the University of New Mexico in Albuquerque, New Mexico. He is a third generation engineer—his grandfather was a civil engineer and father an aeronautical engineer—so "engineering was in my genetic makeup." The genesis of a book on engineering ethics was in the ABET requirement to incorporate ethics topics into the undergraduate engineering curriculum. "Our department decided to have a one-hour seminar course on engineering ethics, but there was no book suitable for such a course." Other texts were tried the first few times the course was offered, but none of them presented ethical theory, analysis, and problem solving in a readily accessible way. "I wanted to have a text which would be concise, yet would give the student the tools required to solve the ethical problems that they might encounter in their professional lives."

http://emissary.prenhall.com/esource/

Reviewers

ESource benefited from a wealth of reviewers who on the series from its initial idea stage to its completion. Reviewers read manuscripts and contributed insightful comments that helped the authors write great books. We would like to thank everyone who helped us with this project.

Concept Document

Naeem Abdurrahman- University of Texas, Austin
Grant Baker- University of Alaska, Anchorage
Betty Barr- University of Houston
William Beckwith- Clemson University
Ramzi Bualuan- University of Notre Dame
Dale Calkins- University of Washington
Arthur Clausing- University of Illinois at Urbana-Champaign
John Glover- University of Houston
A.S. Hodel- Auburn University
Denise Jackson- University of Tennessee, Knoxville
Kathleen Kitto- Western Washington University
Terry Kohutek- Texas A&M University
Larry Richards- University of Virginia
Avi Singhal- Arizona State University
Joseph Wujek- University of California, Berkeley
Mandochehr Zoghi- University of Dayton

Books

Stephen Allan- Utah State University
Naeem Abdurrahman - University of Texas Austin
Anil Bajaj- Purdue University
Grant Baker - University of Alaska - Anchorage
Betty Barr - University of Houston
William Beckwith - Clemson University
Haym Benaroya- Rutgers University
Tom Bledsaw- ITT Technical Institute
Tom Bryson- University of Missouri, Rolla
Ramzi Bualuan - University of Notre Dame
Dan Budny- Purdue University
Dale Calkins - University of Washington
Arthur Clausing - University of Illinois
James Devine- University of South Florida
Patrick Fitzhorn - Colorado State University
Dale Elifrits- University of Missouri, Rolla
Frank Gerlitz - Washtenaw College
John Glover - University of Houston
John Graham - University of North Carolina-Charlotte
Malcom Heimer - Florida International University
A.S. Hodel - Auburn University
Vern Johnson- University of Arizona
Kathleen Kitto - Western Washington University
Robert Montgomery- Purdue University
Mark Nagurka- Marquette University
Ramarathnam Narasimhan- University of Miami
Larry Richards - University of Virginia
Marc H. Richman - Brown University
Avi Singhal-Arizona State University
Tim Sykes- Houston Community College
Thomas Hill- SUNY at Buffalo
Michael S. Wells - Tennessee Tech University
Joseph Wujek - University of California - Berkeley
Edward Young- University of South Carolina
Mandochehr Zoghi - University of Dayton

Contents

1

Introduction

On August 10, 1978, a Ford Pinto was hit from behind on a highway in Indiana. The impact of the collision caused the Pinto's fuel tank to rupture and burst into flames, leading to the deaths of three teenage girls riding in the car. This was not the first time that a Pinto had caught on fire as the result of a rear-end collision. In the seven years since the introduction of the Pinto, there had been some 50 lawsuits related to rear-end collisions. However, this time Ford was charged in a criminal court for the deaths of the passengers.

This case was a significant departure from the norm and had important implications for the Ford engineers and managers. A civil lawsuit could only result in Ford being required to pay damages to the victim's estates. A criminal proceeding, on the other hand, would indicate that Ford was grossly negligent in the deaths of the passengers and could result in jail terms for the Ford engineers or managers who worked on the Pinto.

The case against Ford hinged on charges that it was known that the gas-tank design was flawed and was not in line with accepted engineering standards, even though it did meet applicable federal safety standards at the time. During the trial, it was determined that Ford engineers were aware of the dangers of this design, but management,

SECTIONS

- 1.1 Background Ideas
- 1.2 Why Study Engineering Ethics?
- 1.3 Personal vs. Business Ethics
- 1.4 The Origins of Ethical Thought
- 1.5 Ethics and the Law
- 1.6 Ethics Problems Are Like Design Problems
- 1.7 Case Studies

OBJECTIVES

After reading this chapter, you will be able to:

- Know why it is important to study engineering ethics.
- Understand the distinction between business and personal ethics.
- See how ethical problem solving and engineering design are similar.

concerned with getting the Pinto to market rapidly at a price competitive with subcompact cars already introduced or planned by other manufacturers, had constrained the engineers to use this design.

The dilemma faced by the design engineers who worked on the Pinto was to balance the safety of the people who would be riding in the car against the need to produce the Pinto at a price that would be competitive in the market. They had to attempt to balance their duty to the public against their duty to their employer. Ultimately, the attempt by Ford to save a few dollars in manufacturing costs led to the expenditure of millions of dollars in defending lawsuits and payments to victims. Of course, there were also uncountable costs in lost sales due to bad publicity and a public perception that Ford did not engineer its products to be safe.

1.1 BACKGROUND IDEAS

The Pinto case is just one example of the ethical problems faced by engineers in the course of their professional practice. Ethical cases can go far beyond issues of public safety and may involve bribery, fraud, environmental protection, fairness, honesty in research and testing, and conflicts of interest. During their undergraduate education, engineers receive training in basic and engineering sciences, problem-solving methodology, and engineering design, but generally receive little training in business practices, safety, and ethics.

This problem has partly been corrected, as many engineering education programs now have courses in what's called "engineering ethics"; indeed, ABET, the accreditation board for undergraduate engineering programs, has mandated that ethics topics be incorporated into undergraduate engineering curricula. The purpose of this book is to provide a text and a resource for the study of engineering ethics and to help future engineers be prepared for confronting and resolving ethical dilemmas, such as the design of an unsafe product like the Pinto, that they might encounter during their professional careers.

A good place to start a discussion of ethics in engineering is with definitions of ethics and engineering ethics. Ethics is the study of the characteristics of morals. Ethics also deals with the moral choices that are made by each person in his or her relationship with other persons. As engineers, we are concerned with ethics because these definitions apply to all of the choices an individual makes in life, including those made while practicing engineering.

For our purposes, the definition of ethics can be narrowed a little. Engineering ethics is the rules and standards governing the conduct of engineers in their role as professionals. Engineering ethics encompasses the more general definition of ethics, but applies it more specifically to situations involving engineers in their professional lives. Thus, engineering ethics is a body of philosophy indicating the ways that engineers should conduct themselves in their professional capacity.

1.2 WHY STUDY ENGINEERING ETHICS?

Why is it important for engineering students to study engineering ethics? Several notorious cases that have received a great deal of media attention in the past few years have led engineers to gain an increased sense of their professional responsibilities. These cases have led to an awareness of the importance of ethics within the engineering profession as engineers realize how their technical work has far-reaching impacts on society. The work of engineers can affect public health and safety and can influence business practices and even politics.

One result of this increase in awareness is that nearly every major corporation now has an "ethics office" that has the responsibility to ensure that employees have the ability to express their concerns about issues such as safety and corporate business practices in a way that will yield results and won't result in retaliation against the employee. Ethics offices also try to foster an ethical culture that will help to head off ethical problems in a corporation before they start.

The goal of this book and courses in engineering ethics is to sensitize you to important ethical issues before you have to confront them. You will study important cases from the past so that you will know what situations other engineers have faced and will know what to do when similar situations arise in your professional career. Finally, you will learn techniques for analyzing and resolving ethical problems when they arise.

Our goal is frequently summed up using the term "moral autonomy." Moral autonomy is the ability to think critically and independently about moral issues and to apply this moral thinking to situations that arise in the course of professional engineering practice. The goal of this book, then, is to foster the moral autonomy of future engineers.

The question asked at the beginning of this section can also be asked in a slightly different way. Why should a future engineer bother studying ethics at all? After all, at this point in your life, you're already either a good person or a bad person. Good people already know the right thing to do, and bad people aren't going to do the right thing no matter how much ethical training they receive. The answer to this question lies in the nature of the ethical problems that are often encountered by an engineer. In most situations, the correct response is very obvious. For example, it is clear that to knowingly equip the Pinto with wheel lugs made from substandard, weak steel that is susceptible to breaking is unethical and wrong. This action could lead to the loss of a wheel while driving and could cause numerous accidents and put many lives at risk. Of course, such a design decision would also be a commercial disaster for Ford.

However, many times, the ethical problems encountered in engineering practice are very complex and involve conflicting ethical principles. For example, the engineers working on the Pinto were presented with a very clear dilemma. Trade-offs were made so that the Pinto could be successfully marketed at a reasonable price. One of these trade-offs involved the placement of the gas tank, which led to the accident in Indiana. (This case will be presented in more depth in a later section of this book on safety.) So, for the Ford engineers and managers, the question became the following: Where does an engineering team strike the balance between safety and affordability and, simultaneously, the ability of the company to sell the car and make a profit.

These are the types of situations that we will discuss in this book. The goal, then, is not to train you to do the right thing when the ethical choice is obvious and you already know the right thing to do. Rather, the goal is to train you to analyze complex problems and learn to resolve these problems in the most ethical manner.

1.3 PERSONAL VS. BUSINESS ETHICS

In discussing engineering ethics, it is important to make a distinction between personal ethics and professional, or business, ethics, although there isn't always a clear boundary between the two. Personal ethics deals with how we treat others in our day-to-day lives. Many of these principles are applicable to ethical situations that occur in business and engineering. However, professional ethics often involves choices on an organizational level rather than a personal level. Many of the problems will seem different because they involve relationships between two corporations, between a corporation and the government, or between corporations and groups of individuals. Frequently, these types of relationships pose problems that are not encountered in personal ethics.

1.4 THE ORIGINS OF ETHICAL THOUGHT

Before proceeding, it is important to acknowledge in a general way the origins of the ethical philosophies that we will be discussing in this book. The Western ethical thought that is discussed here originated in the philosophy of the ancient Greeks and their predecessors. It has been developed through subsequent centuries by many thinkers in the Judeo–Christian tradition. Interestingly, non-Western cultures have independently developed similar ethical principles.

Although for many individuals, personal ethics are rooted in religious beliefs, this is not true for everyone. Certainly, there are many ethical people who are not religious, and there are numerous examples of nominally religious people who are not ethical. So while the ethical principles that we will discuss come to us filtered through a religious tradition, these principles are now cultural norms in the West, and as such, they are widely accepted regardless of their origin. We won't need to refer explicitly to religion in order to discuss ethics in the engineering profession.

1.5 ETHICS AND THE LAW

We should also mention the role of law in engineering ethics. The practice of engineering and business is governed by many laws on the international, federal, state, and local levels. Many of these laws are based on ethical principles, although many are purely of a practical, rather than a philosophical, nature. There is also a distinction between what is legal and what is ethical. Many things that are legal could be considered unethical. For example, designing a process that releases a known toxic, but unregulated, substance into the environment is probably unethical, although it is legal.

Conversely, just because something is illegal doesn't mean that it is unethical. For example, there might be substances that were once thought to be harmful, but have now been shown to be safe, that you wish to incorporate into a product. If the law has not caught up with the latest scientific findings, it might be illegal to release these substances into the environment, even though there is no ethical problem with doing so.

As an engineer, you are always minimally safe if you follow the requirements of the applicable laws. But in engineering ethics, we seek to go beyond the dictates of the law. Our interest is in areas where ethical principles conflict *and* there is no legal guidance for how to resolve the conflict.

1.6 ETHICS PROBLEMS ARE LIKE DESIGN PROBLEMS

At first, many engineering students find the types of problems and discussions that take place in an engineering ethics class a little alien. The problems are more open ended and are not as susceptible to formulaic answers as are problems typically assigned in other engineering classes. Ethics problems rarely have a "correct" answer that will be arrived at by everyone in the class. Surprisingly, however, the types of problem-solving techniques that we will use in this book and the nature of the answers that result bear a striking resemblance to the most fundamental engineering activity: engineering design.

The essence of engineering practice is the design of products, structures, and processes. The design problem is stated in terms of specifications: A device must be designed that meets criteria for performance, aesthetics, and price. Within the limits of these specifications, there are many "correct" solutions. There will, of course, be some solutions that are better than others in terms of higher performance or lower cost. Frequently, there will be two (or more) designs that are very different, yet perform identically. For example, competing automobile manufacturers may design a car to meet the same market niche, yet each manufacturer's solution to the problem will be somewhat different. In fact, we will see later that although the Pinto was susceptible to explosion after rear-end impact, other similar subcompact automobiles were not. In engineering design, there is no unique correct answer!

Ethical problem solving shares these attributes with engineering design. Although there will be no unique correct solution to most of the problems we will examine, there will be a range of solutions that are clearly right, some of which are better than others. There will also be a range of solutions that are clearly wrong. There are other similarities between engineering ethics and engineering design. Both apply a large body of knowledge to the solution of a problem, and both involve the use of analytical skills. So although the nature of the solutions to the problems in ethics will be different from those in most engineering classes, approaches to the problems and the ultimate solution will be very similar to those in engineering practice.

1.7 CASE STUDIES

Before starting to learn the theoretical ideas regarding engineering ethics and before looking at some interesting real-life cases that will illustrate these ideas, let's begin by looking at a very well-known engineering ethics case: the space shuttle *Challenger* accident. This case is presented in depth following this chapter, but at this point we will look at a brief synopsis of the case to further illustrate the types of ethical issues and questions that arise in the course of engineering practice.

Many readers are already familiar with some aspects of this case. The space shuttle *Challenger* was launched in extremely cold weather. During the launch, an O-ring on one of the solid-propellant boosters, made more brittle by the cold, failed. This failure led to the explosion during liftoff. Engineers who had designed this booster had concerns about launching under these cold conditions and recommended that the launch be delayed, but they were overruled by their management (some of whom were trained as engineers), who didn't feel that there was enough data to support a delay in the launch. The shuttle was launched, resulting in the well-documented accident.

On the surface, there appear to be no engineering ethical issues here to discuss. Rather, it seems to simply be an accident. The engineers properly recommended that there be no launch, but they were overruled by management. In the strictest sense this can be considered an accident—no one wanted the *Challenger* to explode—but there are still many interesting questions that should be asked. When there are safety concerns, what is the engineer's responsibility before the launch decision is made? After the launch decision is made, but before the actual launch, what duty does the engineer have? If the decision doesn't go the engineer's way, should she complain to upper management? Or should she bring the problem to the attention of the press? After the accident has occurred, what are the duties and responsibilities of the engineers? If the launch were successful, but the *post mortem* showed that the O-ring had failed and an accident had very nearly occurred, what would be the engineer's responsibility? Even if an engineer moves into management, should he separate engineering from management decisions?

These types of questions will be the subject of this book. In subsequent chapters, ideas about the nature of the engineering profession, ethical theories, and the application of these theories to situations that are likely to occur in professional practice will be presented. Many other real-life cases taken from newspaper accounts and books will be discussed to examine what engineers should do when confronted with ethically troubling situations. Most of these cases will be *post mortem* examinations of disasters, but two will involve analysis of situations in which disaster was averted when many of the individuals involved made ethically sound choices and cooperated to solve a problem.

A word of warning is necessary before these cases are studied. The cliché "Hindsight is 20/20" will seem very true as we examine many of these cases. When studying a case several years after the fact and knowing the ultimate outcome, it is easy to see what the right decision should have been. Obviously, had NASA owned a crystal ball and been able to predict the future, the *Challenger* would never have been launched. Had Ford known the number of people who would be killed as a result of gas-tank failures in the Pinto and the subsequent financial losses in lawsuits and criminal cases, it would have found a better solution to the problem of gas-tank placement. However, we rarely have such clear predictive abilities and must base decisions on our best guess of what the outcome will be. It will be important in studying the cases presented here to try to look at them from the point of view of the individuals who were involved at the time, using their best judgment about how to proceed, and not to judge the cases solely based on the outcome.

APPLICATION: THE SPACE SHUTTLE *CHALLENGER* ACCIDENT

The explosion of the space shuttle *Challenger* is perhaps the most widely-written about case in engineering ethics because of the extensive media coverage at the time of the accident and also because of the many available government reports and transcripts of congressional hearings regarding the explosion. The case illustrates many important ethical issues that engineers face: What is the proper role of the engineer when safety issues are a concern? Who should have the ultimate decision-making authority to order a launch? Should the ordering of a launch be an engineering or a managerial decision? This case has already been presented briefly, and we will now take a more in-depth look.

Background

The space shuttle was designed to be a reusable launch vehicle. The vehicle consists of an orbiter, which looks much like a medium-sized airliner (minus the engines!), two solid-propellant boosters, and a single liquid-propellant booster. At takeoff, all of the boosters are ignited and lift the orbiter out of the earth's atmosphere. The solid rocket boosters are only used early in the flight and are jettisoned soon after takeoff, parachute back to earth, and are recovered from the ocean. They are subsequently repacked with fuel and are reused. The liquid-propellant booster is used to finish lifting the shuttle into orbit, at which point the booster is jettisoned and burns up during reentry. The liquid booster is the only part of the shuttle vehicle that is not reusable. After completion of the mission, the orbiter uses its limited thrust capabilities to reenter the atmosphere and glides to a landing.

The accident on January 28, 1986 was blamed on a failure of one of the solid rocket boosters. Solid rocket boosters have the advantage that they deliver far more thrust per pound of fuel than do their liquid-fueled counterparts, but have the disadvantage that once the fuel is lit, there is no way to turn the booster off or even to control the amount of thrust produced. In contrast, a liquid-fuel rocket can be controlled by throttling the supply of fuel to the combustion chamber or can be shut off by stopping the flow of fuel entirely.

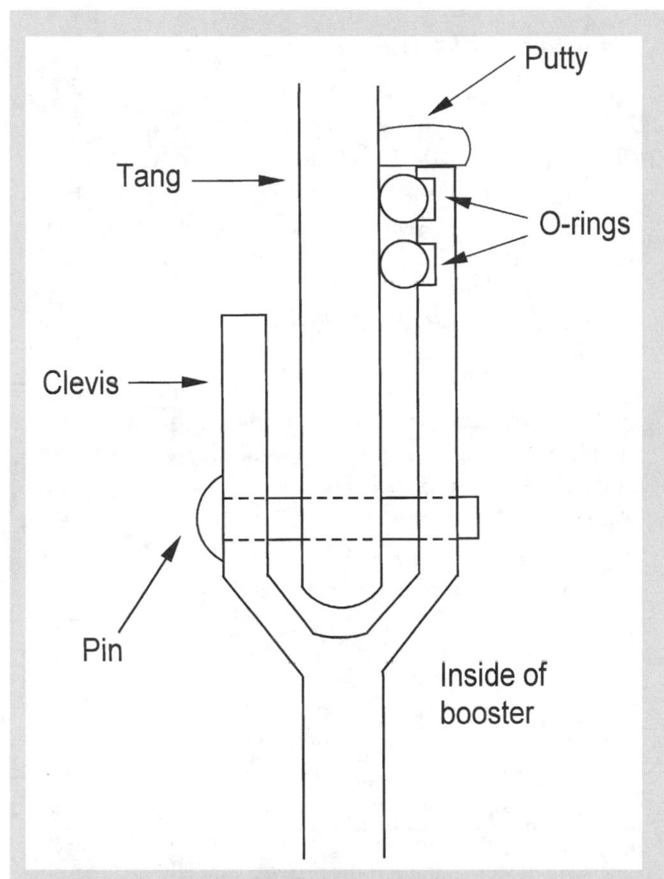

Figure 1.1a. A schematic drawing of a tang and clevis joint like the one on the *Challenger* solid rocket boosters.

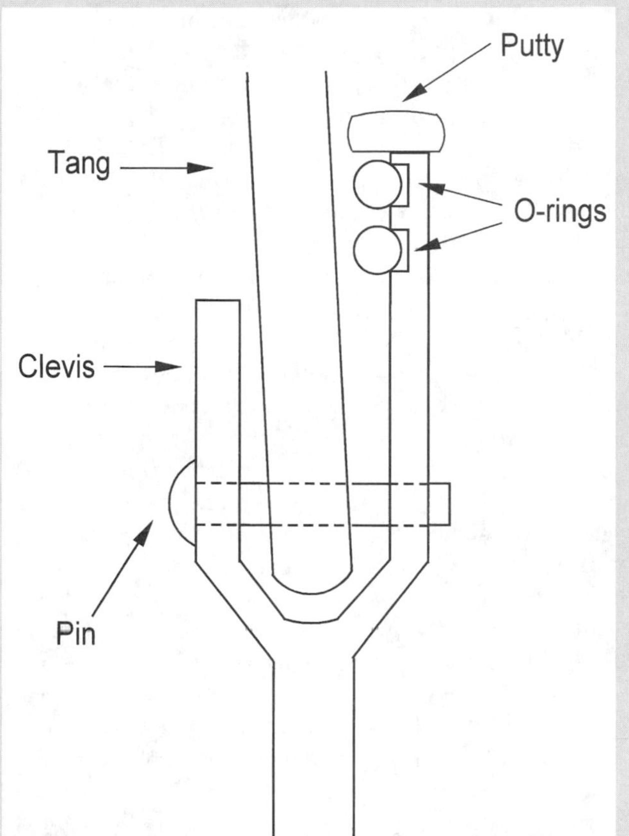

Figure 1.1b. The same joint as in Figure 1a, but with the effects of joint rotation exaggerated. Note that the O-rings no longer seal the joint.

In 1974, the National Aeronautics and Space Administration (NASA) awarded the contract to design and build the solid rocket boosters for the shuttle to Morton Thiokol. The design that was submitted by Thiokol was a scaled-up version of the Titan missile, which had been used successfully for many years to launch satellites. This design was accepted by NASA in 1976. The solid rocket consists of several cylindrical pieces that are filled with solid propellant and stacked one on top of the other to form the completed booster. The assembly of the propellant-filled cylinders was performed at Thiokol's plant in Utah. The cylinders were then shipped to the Kennedy Space Center in Florida for assembly into a completed booster.

A key aspect of the booster design are the joints where the individual cylinders come together, known as the field joints, illustrated schematically in Figure 1a. These are tang and clevis joints, fastened with 177 clevis pins. The joints are sealed by two O-rings, a primary and a secondary. The O-rings are designed to prevent hot gases from the combustion of the solid propellant from escaping. The O-rings are made from a type of synthetic rubber and so are not particularly heat resistant. To prevent the hot gases from damaging the O-rings, a heat-resistant putty is placed in the joint. The Titan booster had only one O-ring in the field joint. The second O-ring was added to the booster for the shuttle to provide an extra margin of safety since, unlike the Titan, this booster would be used for a manned space craft.

Early Problems with the Solid Rocket Boosters

Problems with the field-joint design had been recognized long before the launch of the *Challenger*. When the rocket is ignited, the internal pressure causes the booster wall to expand outward, putting pressure on the field joint. This pressure causes the joint to open slightly, a process called "joint rotation," illustrated in Figure 1b.

The joint was designed so that the internal pressure pushes on the putty, displacing the primary O-ring into this gap, helping to seal it. During testing of the boosters in 1977, Thiokol became aware that this joint-rotation problem was more severe than on the Titan and discussed it with NASA. Design changes were made, including an increase in the thickness of the O-ring, to try to control this problem.

Further testing revealed problems with the secondary seal, and more changes were initiated to correct that problem. In November of 1981, after the second shuttle flight, a postlaunch examination of the booster field joints indicated that the O-rings were being eroded by hot gases during the launch. Although there was no failure of the joint, there was some concern about this situation, and Thiokol looked into the use of different types of putty and alternative methods for applying it to solve the problem. Despite these efforts, approximately half of the shuttle flights before the *Challenger* accident had experienced some degree of O-ring erosion. Of course, this type of testing and redesign is not unusual in engineering. Seldom do things work correctly the first time, and modifications to the original design are often required.

It should be pointed out that erosion of the O-rings is not necessarily a bad thing. Since the solid rocket boosters are only used for the first few minutes of the flight, it might be perfectly acceptable to design a joint in which O-rings erode in a controlled manner. As long as the O-rings don't completely burn through before the solid boosters run out of fuel and are jettisoned, this design should be fine. However, this was not the way the space shuttle was designed, and O-ring erosion was one of the problems that the Thiokol engineers were addressing.

The first documented joint failure came after the launch on January 24, 1985, which occurred during very cold weather. The postflight examination of the boosters revealed black soot and grease on the outside of the booster, which indicated that hot gases from the booster had blown by the O-ring seals. This observation gave rise to concern about the resiliency of the O-ring materials at reduced temperatures. Thiokol performed tests of the ability of the O-rings to compress to fill the joints and found that they were inadequate. In July of 1985, Thiokol engineers redesigned the field joints without O-rings. Instead, they used steel billets, which should have been better able to withstand the hot gases. Unfortunately, the new design was not ready in time for the *Challenger* flight in early 1986. [Elliot, 1991]

The Political Climate

To fully understand and analyze the decision making that took place leading to the fatal launch, it is important also to discuss the political environment under which NASA was operating at the time. NASA's budget was determined by Congress, which was becoming increasingly unhappy with delays in the shuttle project and shuttle performance, which wasn't meeting initial promises. NASA had billed the shuttle as a reliable, inexpensive launch vehicle for a variety of scientific and commercial purposes, including the launching of commercial and military satellites. It had been promised that the shuttle would be capable of frequent flights (several per year) and quick turnarounds and would be competitively priced with more traditional nonreusable launch vehicles. NASA was feeling some urgency in the program because the European Space Agency was developing what seemed to be a cheaper alternative to the shuttle, which could potentially put the shuttle out of business.

These pressures led NASA to schedule a record number of missions for 1986 to prove to Congress that the program was on track. Launching a mission was especially important in January 1986, since the previous mission had been delayed numerous times by both weather and mechanical failures. NASA also felt pressure to get the *Challenger* launched on time so that the next shuttle launch, which was to carry a probe to examine Halley's comet, would be launched before a Russian probe designed to do the same thing. There was additional political pressure to launch the Challenger before the upcoming state-of-the-union address, in which President Reagan hoped to mention the shuttle and a special astronaut—the first teacher in space, Christa McAuliffe—in the context of his comments on education.

The Days Before the Launch

Even before the accident, the *Challenger* launch didn't go off without a hitch, as NASA had hoped. The first launch date had to be abandoned due to a cold front expected to move through the area. The front stalled, and the launch could have taken place on schedule. But the launch had already been postponed in deference to Vice President George Bush, who was to attend. NASA didn't want to antagonize Bush, a strong NASA supporter, by postponing the launch due to inclement weather after he had arrived. Launch of the shuttle was further delayed by a defective microswitch in the hatch-locking mechanism. When this problem

TABLE 1-1 Space Shuttle *Challenger* Accident: Who's Who

ORGANIZATIONS	
NASA	The National Aeronautics and Space Administration, responsible for space exploration. The space shuttle is one of NASA's programs.
Marshall Space Flight Center	A NASA facility that was in charge of the solid rocket booster development for the shuttle.
Morton Thiokol	A private company that won the contract from NASA for building the solid rocket boosters for the shuttle.
PEOPLE	
NASA	
Larry Mulloy	Solid Rocket Booster Project manager at Marshall
Morton Thiokol	
Roger Boisjoly, Arnie Johnson	Engineers who worked on the Solid Rocket Booster Development Program.
Joe Kilminster	Engineering manager on the Solid Rocket Booster Development Program.
Alan McDonald	Director of the Solid Rocket Booster Project.
Bob Lund	Vice president for engineering.
Jerald Mason	General manager.

was resolved, the front had changed course and was now moving through the area. The front was expected to bring extremely cold weather to the launch site, with temperatures predicted to be in the low 20's (°F) by the new launch time.

Given the expected cold temperatures, NASA checked with all of the shuttle contractors to determine if they foresaw any problems with launching the shuttle in cold temperatures. Alan McDonald, the director of Thiokol's Solid Rocket Motor Project, was concerned about the cold weather problems that had been experienced with the solid rocket boosters. The evening before the rescheduled launch, a teleconference was arranged between engineers and management from the Kennedy Space Center, NASA's Marshall Space Flight Center in Huntsville, Alabama, and Thiokol in Utah to discuss the possible effects of cold temperatures on the performance of the solid rocket boosters. During this teleconference, Roger Boisjoly and Arnie Thompson, two Thiokol engineers who had worked on the solid-propellant booster design, gave an hour-long presentation on how the cold weather would increase the problems of joint rotation and sealing of the joint by the O-rings.

Their point was that the lowest temperature at which the shuttle had previously been launched was 53°F, on January 24, 1985, when there was blow-by of the O-rings. The O-ring temperature at *Challenger's* expected launch time the following morning was predicted to be 29°F, far below the temperature at which NASA had previous experience. After the engineer's presentation, Bob Lund, the vice president for engineering at Morton Thiokol, presented his recommendations. He reasoned that since there had previously been severe O-ring erosion at 53°F and the launch would take place at significantly below this temperature where no data and no experience were available, NASA should delay the launch until the O-ring temperature could be at least 53°F. Interestingly, in the original design, it was specified that the booster should operate properly down to an outside temperature of 31°F.

Larry Mulloy, the Solid Rocket Booster Project manager at Marshall and a NASA employee, correctly pointed out that the data were inconclusive and disagreed with the Thiokol engineers. After some discussion, Mulloy asked Joe Kilminster, an engineering manager working on the project, for his opinion. Kilminster backed up the recommendation of his fellow engineers. Others from Marshall expressed their disagreement with the Thiokol engineer's recommendation, which prompted Kilminster to ask to take the discussion off line for a few minutes. Boisjoly and other engineers reiterated to their management that the original decision not to launch was the correct one.

A key fact that ultimately swayed the decision was that in the available data, there seemed to be no correlation between temperature and the degree to which blow-by gasses had eroded the O-rings in previous launches. Thus, it could be concluded that there was really no trend in the data indicating that launch at the expected temperature would necessarily be unsafe. After much discussion, Jerald Mason, a senior manager with Thiokol, turned to Lund and said, "Take off your engineering hat and put on your management hat," a phrase that has become famous in engineering ethics discussions. Lund reversed his previous decision and recommended that the launch proceed. The new recommendation included an indication that there was a safety concern due to the cold weather, but that the data was inconclusive and the launch was recommended. McDonald, who was in Florida, was surprised by this recommendation and attempted to convince NASA to delay the launch, but to no avail.

The Launch

Contrary to the weather predictions, the overnight temperature was 8°F, colder than the shuttle had ever experienced before. In fact, there was a significant accumulation of ice on the launchpad from safety showers and fire hoses that had been left on to prevent the pipes from freezing. It has been estimated that the aft field joint of the right-hand booster was at 28°F.

NASA routinely documents as many aspects of launches as possible. One part of this monitoring is the extensive use of cameras focused on critical areas of the launch vehicle. One of these cameras, looking at the right booster, recorded puffs of smoke coming from the aft field joint immediately after the boosters were ignited. This smoke is thought to have been caused by the steel cylinder of this segment of the booster expanding outward and causing the field joint to rotate. But, due to the extremely cold temperature, the O-ring didn't seat properly. The heat-resistant putty was also so cold that it didn't protect the O-rings, and hot gases burned past both O-rings. It was later determined that this blow-by occurred over 70° of arc around the O-rings.

Very quickly, the field joint was sealed again by byproducts of the solid rocket-propellant combustion, which formed a glassy oxide on the joint. This oxide formation might have averted the disaster had it not been for a very strong wind shear that the shuttle encountered almost one minute into the flight. The oxides that were temporarily sealing the field joint were shattered by the stresses caused by the wind shear. The joint was now opened again, and hot gases escaped from the solid booster. Since the booster was attached to the large liquid-fuel booster, the flames from the solid-fuel booster blow-by quickly burned through the external tank. The liquid propellant was ignited and the shuttle exploded.

The Aftermath

As a result of the explosion, the shuttle program was grounded as a thorough review of shuttle safety was conducted. Thiokol formed a failure-investigation team on January 31, 1986 which included Roger Boisjoly. There were also many investigations into the cause of the accident, both by the contractors involved (including Thiokol) and by various government bodies. As part of the governmental investigation, President Reagan appointed a "blue-ribbon" commission, known as the Rogers commission, after its chair. The commission consisted of distinguished scientists and engineers who were asked to look into the cause of the accident and to recommend changes in the shuttle program.

One of the commission members was Richard Feynman, a Nobel prize winner in physics, who ably demonstrated to the country what had gone wrong. In a demonstration that was repeatedly shown on national news programs, he demonstrated the problem with the O-rings by taking a sample of the O-ring material and bending it. The flexibility of the material at room temperature was evident. He then immersed it in ice water. When Feynman again bent the O-ring, it was very clear that the resiliency of the material was severely reduced, a very clear demonstration of what happened to the O-rings on the cold launch date in Florida.

As part of the commission hearings, Boisjoly and other Thiokol engineers were asked to testify. Boisjoly handed over to the commission copies of internal Thiokol memos and reports detailing the design process and the problems that had already been encountered. Naturally, Thiokol was trying to put the best possible spin on the situation, and Boisjoly's actions hurt this effort. According to Boisjoly, after this action he was isolated within the company, his responsibilities for the redesign of the joint were taken away, and he was subtly harassed by Thiokol management [Boisjoly, 1991, and Boisjoly, Curtis, and Mellicam, 1989].

Explosion of the space shuttle *Challenger* soon after lift-off in January 1986.
NASA/Johnson Space Center.

Eventually, the atmosphere became intolerable for Boisjoly, and he took extended sick leave from his position at Thiokol. The joint was redesigned, and the shuttle has since flown numerous successful missions. However, the ambitious launch schedule originally intended by NASA has never been met.

SUMMARY

Engineering ethics is the study of moral decisions that must be made by engineers in the course of engineering practice. It is important for engineering students to study ethics so that they will be prepared to respond appropriately to ethical challenges during their careers. Often, the correct answer to an ethical problem will not be obvious and will require some analysis using ethical theories. The types of problems that we will encounter in studying engineering ethics are very similar to the design problems that engineers work on every day. As in design, there will not be a single correct answer. Rather, engineering ethics problems will have multiple correct solutions, with some solutions being better than others.

REFERENCES

ROGER BOISJOLY, "The Challenger Disaster: Moral Responsibility and the Working Engineer," in Deborah G. Johnson, *Ethical Issues in Engineering*, Prentice Hall, 1991, pp. 6–14.

NORBERT ELLIOT, ERIC KATZ, and ROBERT LYNCH, "The Challenger Tragedy: A Case Study in Organizational Communication and Professional Ethics," *Business and Professional Ethics Journal*, vol. 12, 1990, pp. 91–108.

JOSEPH R. HERKERT, "Management's Hat Trick: Misuse of 'Engineering Judgment' in the Challenger Incident," *Journal of Business Ethics*, vol. 10, 1991, pp. 617–620.

PATRICIA H. WERHANE, "Engineers and Management: The Challenge of the Challenger Incident," *Journal of Business Ethics*, vol. 10, 1991, pp. 605–16.

ROGER P. BOISJOLY, ELLEN FOSTER CURTIS, and EUGENE MELLICAN, "Roger Boisjoly and the Challenger Disaster: The Ethical Dimensions," *Journal of Business Ethics*, vol. 8, 1989, pp. 217–230.

Problems

1. How different are personal ethics and business ethics? Is this difference true for you personally?

2. What are the roots of your personal ethics?

3. Engineering design generally involves five steps: developing a statement of the problem and/or a set of specifications, gathering information pertinent to the problem, designing several alternatives that meet the specifications, analyzing the alternatives and selecting the best one, and testing and implementing the best design. How is ethical problem solving like this?

Space Shuttle Challenger

4. The astronauts on the *Challenger* mission were aware of the dangerous nature of riding a complex machine such as the space shuttle, so they can be thought of as having given informed consent to participating in a dangerous enterprise. What role did informed consent play in this case? Do you think that the astronauts had enough information to give informed consent to launch the shuttle that day?

5. Can an engineer who has become a manager truly ever take off her engineer's hat? Should she?

6. Some say that the shuttle was really designed by Congress rather than NASA. What does this statement mean? What are the ramifications if this is true?

7. Aboard the shuttle for this flight was the first teacher in space. Should civilians be allowed on what is basically an experimental launch vehicle? At the time, many felt that the placement of a teacher on the shuttle was for purely political purposes. President Reagan was widely seen as doing nothing while the American educational system decayed. Cynics felt that the teacher-in-space idea was cooked up as a method of diverting attention from this problem and was to be seen as Reagan's doing something for education while he really wasn't doing anything. What are the ethical implications if this scenario is true?

8. Should a launch have been allowed when there was no test data for the expected conditions? Keep in mind that it is probably impossible to test for all possible operating conditions. More generally, should a product be released for use even when it hasn't been tested over all expected operational conditions? When the data is inconclusive, which way should the decision go?

9. During the aftermath of the accident, Thiokol and NASA investigated possible causes of the explosion. Boisjoly accused Thiokol and NASA of intentionally downplaying the problems with the O-rings while looking for other causes of the accident. If true, what are the ethical implications of this type of investigation?

10. It might be assumed that the management decision to launch was prompted in part by concerns for the health of the company and the space program as a whole. Given the political climate at the time of the launch, if problems and delays continued, ultimately Thiokol might have lost NASA contracts, or NASA budgets might have been severely reduced. Clearly, this scenario could have lead to the loss of many jobs at Thiokol and NASA. How might these considerations ethically be factored into the decision?

11. Engineering codes of ethics require engineers to protect the safety and health of the public in the course of their duties. Do the astronauts count as "public" in this context?

12. What should NASA management have done differently? What should Thiokol management have done differently?

13. What else could Boisjoly and the other engineers at Thiokol have done to prevent the launch from occurring?

2

Professionalism and Codes of Ethics

Late in 1994, reports began to appear in the news media that the latest generation of Pentium microprocessors, the heart and soul of personal computers, was flawed. These reports appeared not only in trade journals and magazines aimed at computer specialists, but also in *The New York Times* and other daily newspapers. The stories reported that computers equipped with these chips were unable to correctly perform some relatively simple multiplication and division operations.

At first, Intel, the manufacturer of the Pentium microprocessor, denied that there was a problem. Later, it argued that although there was a problem, the error would be significant only in sophisticated applications, and most people wouldn't even notice that an error had occurred. It was also reported that Intel had been aware of the problem and already was working to fix it. As a result of this publicity, many people who had purchased Pentium-based computers asked to have the defective chip replaced. Until the public outcry had reached huge proportions, Intel refused to replace the chips. Finally, when it was clear that this situation was a public relations disaster for them, Intel agreed to replace the defective chips when customers requested.

SECTIONS

OBJECTIVES

After reading this chapter, you will be able to:

- Determine whether engineering is a profession.
- Understand what codes of ethics are.
- Examine some codes of ethics of professional engineering societies.

Did Intel do anything unethical? To answer this question, we will need to develop a framework for understanding ethical problems. One part of this framework will be the codes of ethics that have been established by professional engineering organizations. These codes help guide engineers in the course of their professional duties and give them insight into ethical problems such as the one just described. The engineering codes of ethics hold that engineers should not make false claims or represent a product to be something that it is not. In some ways, the Pentium case might seem to simply be a public-relations problem. But, looking at the problem with a code of ethics will indicate that there is more to this situation than simple PR, especially since the chip did not operate in the way that Intel claimed it did.

In this chapter, the nature of professions will be examined with the goal of determining whether engineering is a profession. Two representative engineering codes of ethics will be looked at in detail. At the end of this chapter, the Pentium case is presented in more detail along with two other cases, and codes of ethics are applied to analyze what the engineers in these cases should have done.

2.1 INTRODUCTION

When confronted by an ethical problem, what resources are available to an engineer to help find a solution? One of the hallmarks of modern professions are codes of ethics promulgated by various professional societies. These codes serve to guide practitioners of the profession in making decisions about how to conduct themselves and how to resolve ethical issues that might confront them. Are codes of ethics applicable to engineering? To answer this question, we must first consider what professions are and how they function and then decide if this definition applies to engineering. Then we will examine codes of ethics in general and look specifically at some of the codes of engineering professional societies.

2.2 IS ENGINEERING A PROFESSION?

In order to determine whether engineering is a profession, the nature of professions must first be examined. As a starting point, it will be valuable to distinguish the word "profession" from other words that are sometimes used synonymously with "profession": "job" and "occupation." Any work for hire can be considered a job, regardless of the skill level involved and the responsibility granted. Engineering is certainly a job—engineers are paid for their services—but the skills and responsibilities involved in engineering make it more than just a job.

Similarly, the word "occupation" implies employment through which someone makes a living. Engineering, then, is also an occupation. How do the words "job" and "occupation" differ from "profession?"

The words "profession" and "professional" have many uses in modern society that go beyond the definition of a job or occupation. One often hears about "professional athletes" or someone referring to himself as a "professional carpenter," for example. In the first case, the word "professional" is being used to distinguish the practitioner from an unpaid amateur. In the second case, it is used to indicate some degree of skill acquired through many years of experience, with an implication that this practitioner will provide quality services.

Neither of these senses of the word "professional" is applicable to engineers. There are no amateur engineers who perform engineering work without being paid while they train to become professional, paid engineers. Likewise, the length of time one works at an engineering-related job, such as an engineering aide or engineering technician, does not confer professional status no matter how skilled a technician one might become. To see what is meant by the term "professional engineer", we will first examine the nature of professions.

What Is a Profession?

What are the attributes of a profession? There have been many studies of this question, and some consensus as to the nature of professions has been achieved. Attributes of a profession include:

1. The work requires sophisticated skills, the use of judgment, and the exercise of discretion. Also, the work is not routine and is not capable of being mechanized;
2. Membership in the profession requires extensive formal education, not simply practical training or apprenticeship;
3. The public allows special societies or organizations that are controlled by members of the profession to set standards for admission to the profession, to set standards of conduct for members, and to enforce these standards; and
4. Significant public good results from the practice of the profession [Martin and Schinzinger, 1989].

The terms "judgment" and "discretion" used in the first part of this definition require a little amplification. Many occupations require judgment every day. A secretary must decide what work to tackle first. An auto mechanic must decide if a part is sufficiently worn to require complete replacement, or if rebuilding will do. This is not the type of judgment implied in this definition. In a profession, "judgment" refers to making significant decisions based on formal training and experience. In general, the decisions will have serious impact on people's lives and will often have important implications regarding the spending of large amounts of money.

"Discretion" can have two different meanings. The first definition involves being discrete in the performance of one's duties by keeping information about customers, clients, and patients confidential. This confidentiality is essential for engendering a trusting relationship and is a hallmark of professions. While many jobs might involve some discretion, this definition implies a high level of significance to the information that must be kept private by a professional. The other definition of discretion involves the ability to make decisions autonomously. When making a decision, one is often told, "Use your discretion." This definition is similar in many ways to that of the term "judgment" described previously. Many people are allowed to use their discretion in making choices while performing their jobs. However, the significance of the decision marks the difference between a job and a profession.

One thing not mentioned in the definition of a profession is the compensation received by a professional for his services. Although most professionals tend to be relatively well compensated, high pay is not a sufficient condition for professional status. Entertainers and athletes are among the most highly paid members of our society, and yet few would describe them as professionals in the sense described previously. Although professional status often helps one to get better pay and better working conditions, these are more often determined by economic forces.

Earlier, reference was made to "professional" athletes and carpenters. Let's examine these occupations in light of the foregoing definition of professions and see if athletics and carpentry qualify as professions. An athlete who is paid for her appearances is referred to as a professional athlete. Clearly, being a paid athlete does involve sophisticated skills that most people do not possess, and these skills are not capable of mechanization. However, substantial judgment and discretion are not called for on the part of athletes in their "professional" lives, so athletics fails the first part of the definition of "professional." Interestingly, though, professional athletes are frequently viewed as role models and are often disciplined for a lack of discretion in their personal lives.

Athletics requires extensive training, not of a formal nature, but more of a practical nature acquired through practice and coaching. No special societies (as opposed to unions, which will be discussed in more detail later) are required by athletes, and athletics does not meet an important public need; although entertainment is a public need, it certainly doesn't rank highly compared to the needs met by professions such as medicine. So, although they are highly trained and very well compensated, athletes are not professionals.

Similarly, carpenters require special skills to perform their jobs, but many aspects of their work can be mechanized, and little judgment or discretion is required. Training in carpentry is not formal, but rather is practical by way of apprenticeships. No organizations or societies are required. However, carpentry certainly does meet an aspect of the public good—providing shelter is fundamental to society—although perhaps not to the same extent as do professions such as medicine. So, carpentry also doesn't meet the basic requirements to be a profession. We can see, then, that many jobs or occupations whose practitioners might be referred to as professionals don't really meet the basic definition of a profession. Although they may be highly paid or important jobs, they are not professions.

Before continuing with an examination of whether engineering is a profession, let's look at two occupations that are definitely regarded by society as professions: medicine and law. Medicine certainly fits the definition of a profession given previously. It requires very sophisticated skills that can't be mechanized, it requires judgment as to appropriate treatment plans for individual patients, and it requires discretion (physicians have even been granted "physician–patient privilege," the duty not to divulge information given in confidence by the patient to the physician). Although medicine requires extensive practical training learned through an apprenticeship called a residency, it also requires much formal training (four years of undergraduate school, three to four years of medical school, and extensive hands-on practice in patient care). Medicine has a special society, the American Medical Association (AMA), to which a large fraction of practicing physicians belong and that participates in the regulation of medical schools, sets standards for practice of the profession, and enforces codes of ethical behavior for its members. Finally, healing the sick and helping to prevent disease clearly involve the public good. By the definition presented previously, medicine clearly qualifies as a profession.

Similarly, law is a profession. It involves sophisticated skills acquired through extensive formal training; has a professional society, the American Bar Association (ABA); and serves an important aspect of the public good (although this last point is increasingly becoming a point of debate within American society!). The difference between athletics and carpentry on one hand and law and medicine on the other is clear. The first two really cannot be considered professions, and the latter two most certainly are.

Engineering as a Profession

Using medicine and law as our examples of professions, it is now time to consider whether engineering is a profession. Certainly, engineering requires extensive and sophisticated skills. Otherwise, why spend four years in college just to get a start in engineering? The essence of engineering design is judgment: how to use the available materials, components, devices, etc. to reach a specified objective. Discretion is required in engineering: Engineers are required to keep their employers' or clients' intellectual-property and business information confidential. Also, a primary concern of any engineer is the safety of the public that will use the products and devices he designs. There is always a trade-off between safety and other engineering issues in a design, requiring discretion on the part of the engineer to ensure that the design serves its purpose and fills its market niche safely.

The point about mechanization needs to be addressed a little more carefully with respect to engineering. Certainly, once a design has been performed, it can easily be replicated without the intervention of an engineer. However, each new situation that requires a new design or a modification of an existing design requires an engineer. Industry commonly uses many computer-based tools for generating designs, such as computer-aided design (CAD) software. This shouldn't be mistaken for mechanization of engineering. CAD is simply a tool used by engineers, not a replacement for the skills of an actual engineer. A wrench can't fix an automobile without a mechanic. Likewise, a computer with CAD software can't design an antilock braking system for an automobile without an engineer.

Engineering requires extensive formal training. Four years of undergraduate training leading to a bachelor's degree in an engineering program is essential, followed by work under the supervision of an experienced engineer. Many engineering jobs even require advanced degrees beyond the bachelor's degree. The work of engineers serves the public good by providing communication systems, transportation, energy resources, and medical diagnostic and treatment equipment, to name only a few.

Before passing final judgment on the professional status of engineering, the nature of engineering societies requires a little consideration. Each discipline within engineering has a professional society, such as the IEEE for electrical engineers and the ASME for mechanical engineers. These societies serve to set professional standards and frequently work with schools of engineering to set standards for admission and curricula. However, these societies differ significantly from the AMA and the ABA. Unlike law and medicine, each specialty of engineering has its own society. There is no overall engineering society that most engineers identify with, although the National Society of Professional Engineers (NSPE) tries to function in this way. In addition, relatively few practicing engineers belong to their professional societies. Thus, the engineering societies are weak compared to the AMA and the ABA.

It is clear that engineering meets all of the definitions of a profession. In addition, it is clear that engineering practice has much in common with medicine and law. Interestingly, although they are professionals, engineers do not yet hold the same status within society that physicians and lawyers do.

Differences Between Engineering and Other Professions

Although we have determined that engineering is a profession, it should be noted that there are significant differences between how engineering is practiced and how law and medicine are practiced. Lawyers are typically self employed in private practice, essentially an independent business, or in larger group practices with other lawyers. Relatively few are employed by large organizations such as corporations. Until recently, this was also the case for most physicians, although with the accelerating trend towards managed care and HMOs in the past decade, many more physicians work for large corporations rather than in private practice. However, even physicians who are employed by large HMOs are members of organizations in which they retain much of the decision-making power—often, the head of an HMO is a physician—and are a substantial fraction of the total number of employees.

In contrast, engineers generally practice their profession very differently from physicians and lawyers. Most engineers are not self-employed, but more often are a small part of larger companies involving many different occupations, including accountants, marketing specialists, and extensive numbers of less skilled manufacturing employees. The exception to this rule is civil engineers, who generally practice as independent consultants either on their own or in engineering firms similar in many ways to law firms. When employed by large corporations, engineers are rarely in significant managerial positions, except with regard to managing other engineers. Although engineers are paid well compared to the rest of society, they are generally less well compensated than physicians and lawyers.

Training for engineers is different than for physicians and lawyers. One can be employed as an engineer after four years of undergraduate education, unlike law and medicine, for which training in the profession doesn't begin until after the undergraduate program has been completed. As mentioned previously, the engineering societies are not as powerful as the AMA and the ABA, perhaps because of the number of different professional engineering societies. Also, both law and medicine require licenses granted by the state in order to practice. Many engineers, especially those employed by large industrial companies, do not have engineering licenses. It can be debated whether someone who is unlicensed is truly an engineer or whether he is practicing engineering illegally, but the reality is that many of those who are employed as engineers are not licensed. Finally, engineering doesn't have the social stature that law and medicine have (a fact that is reflected in the lower pay that engineers receive as compared to that of lawyers and doctors). Despite these differences, on balance, engineering is still clearly a profession, albeit one that is not as mature as medicine and law and that should be striving to emulate some of the aspects of these professions.

Other Aspects of Professional Societies

We should briefly note that professional societies also serve other, perhaps less noble, purposes than those mentioned previously. Sociologists who study the nature of professional societies describe two different models of professions, sometimes referred to as the social-contract and the business models. The social-contract model views professional societies as being set up primarily to further the public good, as described in the definition of a profession given previously. There is an implicit social contract involved with professions, according to this model. Society grants the professions perks such as high pay, a high status in society, and the ability to self-regulate. In return for these perks, society gets the services provided by the profession.

A perhaps more cynical view of professions is provided by the business model. According to this model, professions function as a means for furthering the economic advantage of the members. Put another way, professional organizations are labor unions for the elite, strictly limiting the number of practitioners of the profession, controlling the working conditions for professionals, and artificially inflating the salaries of its members. An analysis of both models in terms of law and medicine would show that there are ways in which these professions exhibit aspects of both of these models.

Where does engineering fit into this picture? Engineering is certainly a service-oriented profession and thus fits into the social-contract model quite nicely. Although some engineers might wish to see engineering professional societies function more according to the business model, they currently don't function that way. The engineering societies have virtually no clout with major engineering employers to set wages and working conditions or to help engineers resolve ethical disputes with their employers.

Moreover, there is very little prospect that the engineering societies will function this way in the near future.

If Engineering Were Practiced More Like Medicine

It is perhaps instructive to speculate a little on how engineering might change in the future if our model of the engineering profession were closer to that of law or medicine. One major change would be in the way engineers are educated. Rather than the current system, in which students study engineering as undergraduates and then pursue advanced degrees as appropriate, prospective engineers would probably get a four-year "preengineering" degree in mathematics, physics, chemistry, computer science, or some combination of these fields. After the four-year undergraduate program, students would enter a three- or four-year engineering professional program culminating in a "doctor of engineering" degree (or other appropriately named degree). This program would include extensive study of engineering fundamentals, specialization in a field of study, and perhaps "clinical" training under a practicing engineer.

How would such engineers be employed? The pattern of employment would certainly be different. Engineers in all fields might work for engineering firms similar to the way in which civil engineers work now, consulting on projects for government agencies or large corporations. The corporate employers who now have numerous engineers on their staff would probably have far fewer engineers on the payroll, opting instead for a few professional engineers who would supervise the work of several less highly trained "engineering technicians." Adoption of this model would probably reduce the number of engineers in the work force, leading to higher earnings for those who remain. Those relegated to the ranks of engineering technicians would probably earn less than those currently employed as engineers.

2.3 CODES OF ETHICS

An aspect of professional societies that has not been mentioned yet is the codes of ethics that engineering societies have adopted. These codes express the rights, duties, and obligations of the members of the profession. In this section, we will examine the codes of ethics of professional engineering societies, several of which are included in Appendix A.

It should be noted that although most of the discussion thus far has focused on professionalism and professional societies, codes of ethics are not limited to professional organizations. They can also be found, for example, in corporations and universities as well. We start with some general ideas about what codes of ethics are and what purpose they serve and then examine two professional engineering codes in more detail.

What Is a Code of Ethics?

Primarily, a code of ethics provides a framework for ethical judgment for a professional. The key word here is "framework." No code can be totally comprehensive and cover all possible ethical situations that a professional engineer is likely to encounter. Rather, codes serve as a starting point for ethical decision making. A code can also express the commitment to ethical conduct shared by members of a profession. It is important to note that ethical codes do not establish new ethical principles. They simply reiterate principles and standards that are already accepted as responsible engineering practice. A code expresses these principles in a coherent, comprehensive, and accessible manner. Finally, a code defines the roles and responsibilities of professionals [Harris, Pritchard, and Rabins, 1995].

It is important also to look at what a code of ethics is not. It is not a recipe for ethical behavior; as previously stated, it is only a framework for arriving at good ethical choices. A code of ethics is never a substitute for sound judgment. A code of ethics is not a legal document. One can't be arrested for violating its provisions, although expulsion from the professional society might result from code violations. As mentioned in the previous section, with the current state of engineering societies, expulsion from an engineering society generally will not result in an inability to practice engineering, so there are not necessarily any direct consequences of violating engineering ethical codes. Finally, a code of ethics doesn't create new moral or ethical principles. As described in the previous chapter, these principles are well established in society, and foundations of our ethical and moral principles go back many centuries. Rather, a code of ethics spells out the ways in which moral and ethical principles apply to professional practice. Put another way, a code helps the engineer to apply moral principles to the unique situations encountered in professional practice.

How does a code of ethics achieve these goals? First, a code of ethics helps create an environment within a profession where ethical behavior is the norm. It also serves as a guide or reminder of how to act in specific situations. A code of ethics can also be used to bolster an individual's position with regard to a certain activity: The code provides a little backup for an individual who is being pressured by a superior to behave unethically. A code of ethics can also bolster the individual's position by indicating that there is a collective sense of correct behavior; there is strength in numbers. Finally, a code of ethics can indicate to others that the profession is seriously concerned about responsible, professional conduct [Harris, Pritchard, and Rabins, 1995]. A code of ethics, however, should not be used as "window dressing," an attempt by an organization to appear to be committed to ethical behavior when it really is not.

Objections to Codes

Although codes of ethics are widely used by many organizations, including engineering societies, there are many objections to codes of ethics, specifically as they apply to engineering practice. First, as mentioned previously, relatively few practicing engineers are members of professional societies and so don't necessarily feel compelled to abide by their codes. Many engineers who are members of professional societies are not aware of the existence of the society's code, or if they are aware of it, they have never read it. Even among engineers who know about their society's code, consultation of the code is rare. There are also objections that the engineering codes often have internal conflicts, but don't give a method for resolving the conflict. Finally, codes can be coercive: They foster ethical behavior with a stick rather than with a carrot [Harris, Pritchard, and Rabins, 1995]. Despite these objections, codes are in very widespread use today and are generally thought to serve a useful function.

Codes of the Engineering Societies

Before examining two representative professional codes from Appendix A in more detail, it might be instructive to look briefly at the history of the engineering codes of ethics. Professional engineering societies in the United States began to be organized in the late 19th century, with new societies created as new engineering fields have developed in this century. As these societies matured, many of them created codes of ethics to guide practicing engineers.

Early in the current century, these codes were mostly concerned with issues of how to conduct business. For example, many early codes had clauses forbidding advertising of services or prohibiting competitive bidding by engineers for design projects. Codes also spelled out the duties that engineers had toward their employers. Relatively less emphasis than today was given to issues of service to the public and safety. This imbalance changed greatly in recent decades as public perceptions and concerns about the safety of engineered products and devices have changed. Now, most codes emphasize commitments to safety, public health, and even environmental protection as the most important duties of the engineer.

A Closer Look at Two Codes of Ethics

Having looked at some ideas about what codes of ethics are and how they function, let's look more closely at two of the codes of ethics reproduced in Appendix A: the codes of the Institute of Electrical and Electronics Engineers (IEEE) and the National Society of Professional Engineers (NSPE). Although these codes have some common content, the structures of the codes are very different.

The IEEE code is short and deals in generalities, whereas the NSPE code is much longer and more detailed. An explanation of these differences is rooted in the philosophy of the authors of these codes. A short code that is lacking in detail is more likely to be read by members of the society than is a longer code. A short code is also more understandable. It articulates general principles and truly functions as a framework for ethical decision making, as described previously.

A longer code, such as the NSPE code, has the advantage of being more explicit and able to cover more ground. It leaves less to the imagination of the individual and therefore is more useful for application to specific cases. The length of the code, however, makes it less likely to be read and thoroughly understood by most engineers.

There are some specifics of these two codes that are worth noting here. The IEEE code doesn't mention a duty to one's employer. However, the IEEE code does mention a duty to protect the environment, a clause added relatively recently, which is somewhat unique among engineering codes. The NSPE code has a preamble that succinctly presents the duties of the engineer before going on to the more explicit discussions of the rest of the code. Like most codes of ethics, the NSPE code does mention the engineer's duty to her employer in Section I.4, where it states that engineers shall "[a]ct . . . for each employer . . . as faithful agents or trustees."

Resolving Internal Conflicts in Codes

One of the objections to codes of ethics mentioned previously is the internal conflicts that can exist within them, with no instructions on how to resolve these conflicts. An example of this problem would be a situation in which an employer asks or even orders an engineer to implement a design that the engineer feels will be unsafe. It is made clear that the engineer's job is at stake if he doesn't do as instructed. What does the NSPE code tell us about this situation?

In clause I.4, the NSPE code indicates that engineers have a duty to their employers, which implies that the engineer should go ahead with the unsafe design favored by his employer. However, clause I.1 and the preamble make it clear that the safety of the public is also an important concern of an engineer. In fact, it says that the safety of the public is paramount. How can this conflict be resolved?

There is no implication in this or any other code that all clauses are equally important. Rather, there is a hierarchy within the code. Some clauses take precedence over others, although there is generally no explicit indication in the code of what the hierarchy is. The dilemma presented above is easily resolved within the context of this hierarchy. The duty to protect the safety of the public is paramount and takes precedence over the duty to the employer. In this case, the code provides very clear support to the engineer, who must convince his supervisor that he can't design the product as requested. Unfortunately, not all internal conflicts in codes of ethics are so easily resolved.

Can Codes and Professional Societies Protect Employees?

One important area where professional societies can and should function is as protectors of the rights of employees who are being pressured by their employer to do something unethical or who are accusing their employers or the government of unethical conduct. The codes of the professional societies are of some use in this since they can be used by employees as ammunition against an employer who is sanctioning them for pointing out unethical behavior or who are being asked to engage in unethical acts.

An example of this situation, which we shall discuss in more detail in a later chapter, is the action of the IEEE on behalf of three electrical engineers who were fired from their jobs at the Bay Area Rapid Transit (BART) organization when they pointed out deficiencies in the way the control systems for the BART trains were being designed and tested. After being fired, the engineers sued BART, citing the IEEE code of ethics which impelled them to hold as their primary concern the safety of the public who would be using the BART system. The IEEE intervened on their behalf in court, although ultimately the engineers lost the case.

If the codes of ethics of professional societies are to have any meaning, this type of intervention is essential when ethical violations are pointed out. However, as mentioned previously, since not all engineers are members of professional societies and the engineering societies are relatively weak, the pressure that can be exerted by these organizations is limited.

Other Types of Codes of Ethics

Professional societies aren't the only organizations that have codified their ethical standards. Many other organizations have also developed codes of ethics for various purposes similar to those of the professional engineering organizations. For example, codes for the ethical use of computers have been developed, and student organizations in universities have framed student codes of ethics. In this section, we will examine how codes of ethics function in corporations.

Many of the important ethical questions faced by engineers come up in the context of their work for corporations. Since most practicing engineers are not members of professional organizations, it seems that for many engineers, there is little ethical guidance in the course of their daily work. This problem has led to the adoption of codes of ethics by many corporations.

Even if the professional codes were widely adopted and recognized by practicing engineers, there would still be some value to the corporate codes, since a corporation can tailor its code to the individual circumstances and unique mission of the company. As such, these codes tend to be relatively long and very detailed, incorporating many rules specific to the practices of the company. For example, corporate codes frequently spell out in detail the company policies on business practices, relationships with suppliers, relationships with government agencies, compliance with government regulations, health and safety issues, issues related to environmental protection, equal employment opportunity and affirmative action, sexual harassment, and diversity and racial/ethnic tolerance. Since corporate codes are coercive in nature—your continued employment by the company depends on your compliance with the company code—these codes tend to be longer and more detailed in order to provide very clear and specific guidelines to the employees.

Codes of professional societies, by their nature, can't be this explicit, since there is no means for a professional society to reasonably enforce its code. Due to the typically long lengths of these codes, no example of a corporate code of ethics can be included in

Appendix A with the other types of codes. However, codes for companies can sometimes be found via the Internet at corporate Web sites.

Some of the heightened awareness of ethics in corporations stems from the increasing public scrutiny that has accompanied well-publicized disasters, such as the cases presented elsewhere in this book, as well as from cases of fraud and cost overruns, particularly in the defense industry, that have been exposed in the media. Many large corporations have developed corporate codes of ethics in response to these problems, to help heighten employee's awareness of ethical issues, and to help establish a strong corporate ethics culture. These codes give employees ready access to guidelines and policies of the corporations. But, as with professional codes, it is important to remember that these codes cannot cover all possible situations that an employee might encounter; there is no substitute for good judgment. A code also doesn't substitute for good lines of communications between employees and upper management and for workable methods for fixing ethical problems when they occur.

APPLICATION: CASES

Codes of ethics can be used as a tool for analyzing cases and for gaining some insight into the proper course of action. Before reading these cases, it would be helpful to read a couple of the codes in Appendix A, especially the code most closely related to your field of study, to become familiar with the types of issues that codes deal with. Then, put yourself in the position of an engineer working for these companies—Intel, Paradyne computers, and 3Bs construction—to see what you would have done in each case.

The Intel Pentium Chip

In late 1994, the media began to report that there was a flaw in the new Pentium microprocessor produced by Intel. The microprocessor is the heart of a personal computer and controls all of the operations and calculations that take place. A flaw in the Pentium was especially significant, since it was the microprocessor used in 80% of the personal computers produced in the world at that time.

Apparently, flaws in a complicated integrated circuit such as the Pentium, which at the time contained over one million transistors, are common. However, most of the flaws are undetectable by the user and don't affect the operation of the computer. Many of these flaws are easily compensated for through software. The flaw that came to light in 1994 was different: It was detectable by the user. This particular flaw was in the floating-point unit (FPU) and caused a wrong answer when double-precision arithmetic, a very common operation, was performed.

A standard test was widely published to determine whether a user's microprocessor was flawed. Using spreadsheet software, the user was to take the number 4,195,835, multiply it by 3,145,727, and then divide that result by 3,145,727. As we all know from elementary math, when a number is multiplied and then divided by the same number, the result should be the original number. In this example, the result should be 4,195,835. However, with the flawed FPU, the result of this calculation was 4,195,579. [Infoworld, 1994] Depending on the application, this six-thousandths-of-a-percent error might be very significant.

At first, Intel's response to these reports was to deny that there was any problem with the chip. When it became clear that this assertion was not accurate, Intel switched its policy and stated that although there was indeed a defect in the chip, it was insignificant and the vast majority of users would never even notice it. The chip would be replaced for free only for users who could demonstrate that they needed an unflawed version of the chip. [Infoworld, 1994] There is some logic to this policy from Intel's point of view, since over two million computers had already been sold with the defective chip.

Of course, this approach didn't satisfy most Pentium owners. After all, how can you predict whether you might have a future application where this flaw might be significant? IBM, a major Pentium user, canceled the sales of all IBM computers containing the flawed chip. Finally, after much negative publicity in the popular personal-computer literature and an outcry from Pentium users, Intel agreed to replace the flawed chip with an unflawed version for any customer who asked to have it replaced.

It should be noted that long before news of the flaw surfaced in the popular press, Intel was aware of the problem and had already corrected it on subsequent versions. It did, however, continue to sell the flawed version and, based on its early insistence that the flaw did not present a significant problem to users, seemingly planned to do so until the new version was available and the stocks of the flawed one were exhausted. Eventually, the damage caused by this case was fixed as the media reports of the problem died down and as customers were able to get unflawed chips into their computers.

What did Intel learn from this experience? The early designs for new chips continue to have flaws, and sometimes these flaws are not detected until the product is already in use by consumers. However, Intel's approach to these problems has changed. It now seems to feel that problems need to be fixed immediately. In addition, the decision is now based on the consumer's perception of the significance of the flaw, rather than on Intel's opinion of its significance.

Runway Concrete at the Denver International Airport

In the early 1990s, the city of Denver, Colorado embarked on one of the largest public works projects in history: the construction of a new airport to replace the aging Stapleton International Airport. The new Denver International Airport (DIA) would be the first new airport constructed in the US since the Dallas–Fort Worth Airport was completed in the early 1970s. Of course, the size and complexity of this type of project lends itself to many problems, including cost overruns, worker safety and health issues, and controversies over the need for the project. The construction of DIA was no exception.

Perhaps the most widely known problem with the airport was the malfunctioning of a new computer-controlled high-tech baggage handling system, which in preliminary tests consistently mangled and misrouted baggage and frequently jammed, leading to the shutdown of the entire system. Problems with the baggage handling system delayed the opening of the airport for over a year and cost the city millions of dollars in expenses for replacement of the system and lost revenues while the airport was unable to open. In addition, the baggage system made the airport the butt of many jokes, especially on late-night television. More interesting from the perspective of engineering ethics are problems during the construction of DIA involving the concrete used for the runways, taxiways, and aprons at the airport.

The story of concrete problems at DIA was first reported by the *Denver Post* in early August of 1993 as the airport neared completion. Two subcontractors filed lawsuits against the runway-paving contractor, California construction company Ball, Ball, & Brosamer (known as 3Bs), claiming that 3Bs owed them money. Parts of these suits were allegations that 3Bs had altered the recipe for the concrete used in the runway and apron construction, deliberately diluting the concrete with more gravel, water, and sand (and thus less cement), thereby weakening it. 3Bs motivation for doing so would be to save money and thus to increase their profits. One of the subcontractors, CSI Trucking, whose job was to haul the sand and gravel used in the concrete, claimed that 3Bs hadn't paid them for materials that had been delivered. They claimed that these materials had been used to dilute the mixture, but hadn't been paid for, since the payment would leave a record of the improper recipe.

At first, Denver officials downplayed the reports of defective concrete, relying on the results of independent tests of the concrete. In addition, the city of Denver ordered core samples to be taken from the runways. Tests on these cores showed that the runway concrete had the correct strength. The subcontractors claimed that the improperly mixed concrete could have the proper test strength, but would lead to a severely shortened runway lifetime. The FBI also became involved in investigating this case, since federal transportation grants were used by Denver to help finance the construction of the runways.

The controversy seemed to settle down for a while, but a year later, in August of 1994, the Denver district attorney's office announced that it was investigating allegations that inspection reports on the runways were falsified during the construction. This announcement was followed on November 13, 1994 by a lengthy story in the *Denver Post* detailing a large number of allegations of illegal activities and unethical practices with regard to the runway construction.

The November 13 story revolved around an admission by a Fort Collins, Colorado company, Empire Laboratories, that test reports on the concrete had been falsified to hide results which showed that some of the concrete did not meet the specifications. Attorneys for Empire said that this falsification had happened five or six times in the course of this work, but four employees of Empire claimed that the altering of test data was standard operating procedure at Empire.

The nature of the test modifications and the rationale behind them illustrate many of the important problems we will discuss in this book, including the need for objectivity and honesty in reporting results of tests and experiments. One Empire employee said that if a test result was inconsistent with other tests, then the results would be changed to mask the difference. This practice was justified by Empire as being "based upon engineering judgment" [*Denver Post*, Nov. 13, 1994]. The concrete was tested by pouring test samples when the actual runways were poured. These samples were subjected to flexural tests, which consist of subjecting the concrete to an increasing force until it fails. The tests were performed at 7 days after pouring and also at 28 days. Many of the test results showed that the concrete was weaker at 28 days than at 7 days. However, the results should have been the opposite, since concrete normally increases in strength as it cures. Empire employees indicated that this apparent anomaly was because many of the 7-day tests had been altered to make the concrete seem stronger than it was.

Other problems with the concrete also surfaced. Some of the concrete used in the runways contained clay balls up to 10 inches in diameter. While not uncommon in concrete batching, the presence of this clay can lead to runways that are significantly weaker than planned.

Questions about the short cement content in 3Bs concrete mixture also resurfaced in the November *Denver Post* article. The main question was: Given that the concrete batching operation was routinely monitored, how did 3Bs get away with shorting the cement content of the concrete? One of the batch plant operators for 3Bs explained that they were tipped off about upcoming inspections. When an inspector was due, they used the correct recipe so that concrete would appear to be correctly formulated. The shorting of the concrete mixture could also be detected by looking at the records of materials delivered to the batch plants. However, DIA administrators found that this documentation was missing, and it was unclear whether it had ever existed.

A batch plant operator also gave a sworn statement that he had been directed to fool the computer that operated the batch plant. The computer was fooled by tampering with the scale used to weigh materials and by inputting false numbers for the moisture content of the sand. In some cases, the water content of the sand that was input into the computer was a negative number! This tampering forced the computer to alter the mixture to use less cement, but the records printed by the computer would show that the mix was properly constituted. In this statement, the batch plant operator also swore that this practice was known to some of the highest officials in 3Bs.

Despite the problems with the batching of the concrete used in the runways, DIA officials insisted that the runways built by 3Bs met the specifications. This assertion was based on the test results, which showed that although some parts of the runway were below standard, all of the runways met FAA specifications. 3Bs was paid for those areas that were below standard at a lower rate than for the stronger parts of the runway. Further investigations about misdeeds in the construction of DIA were performed by several groups, including a Denver grand jury, a federal grand jury, the FBI, and committees of Congress.

On October 19, 1995, the *Denver Post* reported the results of a lawsuit brought by 3Bs against the city of Denver. 3Bs contended that the city still owed them $2.3 million (in addition to the $193 million that 3Bs had already been paid) for the work they did. The city claimed that this money was not owed. The reduction was a penalty due to low test results on some of the concrete. 3Bs claimed that those tests were flawed and that the concrete was fine. A hearing officer sided with the city, deciding that Denver didn't owe 3Bs any more money. 3Bs said that they would take their suit to the next higher level.

As of the summer of 1998, DIA has been in operation for over three years and no problems have surfaced regarding the strength of the runways. Unfortunately, problems with runway durability might not surface until after several more years of use. In the meantime, there is still plenty of litigation and investigation of this and other unethical acts surrounding the construction of this airport.

Competitive Bidding and the Paradyne Case

Although competitive bidding is a very well-established practice in purchasing, it can lead to many ethical problems associated with deception on the part of the vendor or with unfairness on the part of the buyer in choosing a vendor. The idea behind competitive bidding is that the buyer can get a product at the best price by setting up competition between the various suppliers. Especially with large contracts, the temptation to cheat on the bidding is great. Newspapers frequently report stories of deliberate underbidding to win contracts, followed by cost overruns that are unavoidable; theft of information on others' bids in order to be able to underbid them; etc. Problems also exist with buyers who make purchase decisions based on elements other than the advertised bid criteria, who leak information to a preferred bidder, or who give advance notice or detailed knowledge of evaluation procedures to preferred bidders. The Paradyne computer case is useful in illustrating some of the hazards associated with competitive bidding.

The Paradyne case began on June 10, 1980, when the Social Security Administration (SSA) published a request for proposals (RFP) for computer systems to replace the older equipment in its field offices. Its requirement was for computers that provide access to a central database. This database was used by field offices in the processing of benefit claims and in issuing new social security numbers. SSA intended to purchase an off-the-shelf system already in the vendor's product line, rather than a customized system. This requirement was intended to minimize the field testing and bugs associated with customized systems. In March of 1981, SSA let a contract for $115 million for 1,800 computer systems to Paradyne.

Problems occurred immediately upon award of the contract, when the Paradyne computers failed the acceptance testing. The requirements were finally relaxed so that the computers would pass. After delivery, many SSA field offices reported frequent malfunctions, sometimes multiple times per day, requiring manual rebooting of the system. One of the contract requirements was that the computers function 98% of the time. This requirement wasn't met until after 21 months of operation. After nearly two years of headaches and much wasted time and money, the system finally worked as planned. [Davis, 1988]

Subsequent investigation by SSA indicated that the product supplied by Paradyne was not an off-the-shelf system, but rather was a system that incorporated new technology that had yet to be built and was still under development. Paradyne had proposed selling SSA their P8400 model with the PIOS operating system. The bid was written as if this system currently existed. However, at the time that the bid was prepared, the 8400 system did not exist and had not been developed, prototyped, or manufactured. [Head, 1986]

There were other problems associated with Paradyne's performance during the bidding. The RFP stated that there was to be a preaward demonstration of the product, not a demonstration of a prototype. Paradyne demonstrated to SSA a different computer, a modified PDP 11/23 computer manufactured by Digital Equipment Corporation (DEC) placed in a cabinet that was labeled P8400. Apparently, many of the DEC labels on the equipment that was demonstrated to SSA had Paradyne labels pasted over them. Paradyne disingenuously claimed that since the DEC equipment was based on a 16-bit processor, as was the P8400 they proposed, it was irrelevant if the machine demonstrated were the DEC or the actual P8400. Of course, computer users recognize that this statement is nonsense. Even modern "PC-compatible" computers with the same microprocessor chip and operating system can have widely different operating characteristics in terms of speed, software that can be run, etc.

There were also questions about the operating system. Apparently, at the time of Paradyne's bid, the PIOS system was under development as well and hadn't been tested on a prototype of the proposed system. Even a functioning hardware system will not operate correctly without the correct operating system. No software has ever worked correctly the first time, but rather requires extensive "debugging" to make it operate properly with a new system. Significantly, the DEC system with the P8400 label that was actually tested by SSA was not running with the proposed PIOS system.

Some of the blame for this fiasco can also be laid at the feet of the SSA. There were six bidders for this contract. Each of the bidders was to have an on-site visit from SSA inspectors to determine whether it was capable of doing the work that it included in its bid. Paradyne's capabilities were not assessed using an on-site visit. Moreover, Paradyne was judged based on its ability to manufacture modems, which was then its main business. Apparently, its ability to produce complete computer systems wasn't assessed. As part of its attempt to gain this contract, Paradyne hired a former SSA official who, while still working for SSA, had participated in preparing the RFP and had helped with setting up the team that would evaluate the bids. Paradyne had notified SSA of the hiring of this person, and

SSA decided that there were no ethical problems with this. However, when the Paradyne machine failed the initial acceptance test, this Paradyne official was directly involved in negotiating the relaxed standards with his former boss at SSA.

This situation was resolved when the Paradyne computers were finally brought to the point of functioning as required. However, as a result of these problems, there were many investigations by government agencies, including the Securities and Exchange Commission, the General Accounting Office, the House of Representative's Government Operations Committee, the Health and Human Services Department (of which SSA is part), and the Justice Department.

KEY TERMS

Professions
Professional societies
Code of ethics

REFERENCES

HARRIS, JR., CHARLES E., MICHAEL S. PRITCHARD, AND MICHAEL J. RABINS, *Engineering Ethics, Concepts and Cases*, Wadsworth Publishing Company, Belmont CA., 1995.

MARTIN, MIKE W., AND ROLAND SCHINZINGER, *Ethics in Engineering*, 2nd edition, McGraw-Hill, New York, 1989.

Intel Pentium Chip Case

"When the Chips are Down," *Time*, Dec. 26–Jan. 2 1995, p. 126.

"The Fallout from Intel's Pentium Bug," *Fortune*, Jan. 16, 1995, p. 15.

"Pentium Woes Continue," *Infoworld*, Nov. 18, 1994, vol. 16, issue 48, p. 1.

"Flawed Chips Still Shipping," *Infoworld*, Dec. 5, 1994, vol. 16, issue 49, p. 1.

Numerous other accounts from late 1994 and early 1995 in The *Wall Street Journal, The New York Times*, etc.

DIA Runaway Concrete

LOU KILZER, ROBERT KOWALSKI, AND STEVEN WILMSEN, "Concrete tests faked at airport," *Denver Post*, Nov. 13, 1994, Section A, p. 1.

Paradyne Computers

J. STEVE DAVIS, "Ethical Problems in Competitive Bidding: The Paradyne Case," *Business and Professional Ethics Journal*, vol. 7, 1988, p. 3.

ROBERT V. HEAD, "Paradyne Dispute: A Matter of Using a Proper Tense," *Government Computer News*, February 14, 1986, p. 23.

Problems

1. What changes would have to be made for engineering to be a profession more like medicine or law?
2. In which ways do law, medicine, and engineering fit the social-contract and the business models of a profession?
3. The first part of the definition of a profession presented previously said that professions involve the use of sophisticated skills. Do you think that these skills are primarily physical or intellectual skills? Give examples from professions such as law, medicine, and engineering, as well as from nonprofessions.
4. Apply one of the codes of ethics in Appendix A to the space shuttle *Challenger* case described at the end of Chapter 1. What guidance might one of the engineering society codes of ethics have given the Thiokol engineers when faced with a decision to launch? Which specific parts of the code are applicable to this situation? Does a manager who is trained as an engineer still have to adhere to an engineering code of ethics?

5. Write a code of ethics for students in your college or department. Start by deciding what type of code you want: short, long, detailed, etc. Then, list the important ethical issues you think students face. Finally, organize these ideas into a coherent structure.
6. Imagine that you are the president of a small high-technology firm. Your company has grown over the last few years to the point where you feel that it is important that your employees have some guidelines regarding ethics. Define the type of company you are running; then develop an appropriate code of ethics. As in Question 2, start by deciding what type of code is appropriate for your company. Then, list specific points that are important—for example, relationships with vendors, treatment of fellow employees, etc. Finally, write a code that incorporates these features.

Intel Pentium Chip

7. Was this case simply a customer-relations and PR problem, or are there ethical issues to be considered as well?
8. Use one of the codes of ethics in Appendix A to analyze this case. Especially, pay attention to issues of accurate representation of engineered products and to safety issues.
9. When a product is sold, is there an implication that it will work as advertised?
10. Should you reveal defects in a product to a consumer? Is the answer to this question different if the defect is a safety issue rather than simply a flaw? (It might be useful to note in this discussion that although there is no apparent safety concern for someone using a computer with this flaw, PCs are often used to control a variety of instruments, such as medical equipment. For such equipment, a flaw might have a very real safety implication.) Is the answer to this question different if the customer is a bank that uses the computer to calculate interest paid, loan payments, etc. for customers?
11. Should you replace defective products even if customers won't recognize the defect?
12. How thorough should testing be? Is it ever possible to say that no defect exists in a product or structure?
13. Do flaws that Intel found previously in the 386 and 486 chips have any bearing on these questions? In other words, if Intel got away with selling flawed chips before without informing consumers, does that fact have any bearing on this case?
14. G. Richard Thoman, an IBM senior vice president was quoted as saying, "Nobody should have to worry about the integrity of data calculated on an IBM machine." How does this statement by a major Intel customer change the answers to the previous questions?
15. Just prior to when this problem surfaced, Intel had begun a major advertising campaign to make Intel a household name. They had gotten computer manufacturers to place "Intel Inside" labels on their computers and had spent money on television advertising seeking to increase the public demand for computers with Intel processors, with the unstated message that Intel chips were of significantly higher quality than other manufacturers' chips. How might this campaign have affected what happened in this case?
16. What responsibilities did the engineers who were aware of the flaw have before the chip was sold? After the chips began to be sold? After the flaw became apparent?

DIA Runaway Concrete

17. Using one of the codes of ethics in Appendix A, analyze the actions of the batch plant operators and Empire Laboratories.
18. Is altering data a proper use of "engineering judgment"? What alternative might have existed to altering the test data on the concrete?
19. Who is responsible for ensuring that the materials used in a project meet the specifications, the supplier or the purchaser?

Paradyne Computers

20. Choose one of the codes of ethics from Appendix A and use it to analyze this case. Were the engineers and managers of Paradyne operating ethically?
21. In preparing their bid, Paradyne wrote in the present tense, as if the computer they proposed currently existed, rather than in the future tense, which would have indicated that the

product was still under development. Paradyne claimed that the use of the present tense in its bid (which led SSA to believe that the P8400 actually existed) was acceptable, since it is common business practice to advertise products under development this way. Was this a new product announcement with a specified availability date? Is there a distinction between a response to a bid and company advertising? Is it acceptable to respond to a bid with a planned system if there is no indication when that system is expected to be available?

22. Paradyne also claimed that it was acting as a system integrator (which was allowed by the RFP), using components from other manufacturers to form the Paradyne system. These other components were mostly off-the-shelf, but they had never been integrated grated into a system before. Does this meet the SSA requirement for an existing system?
23. Once the Paradyne machine failed the initial test, should the requirements have been relaxed to help the machine qualify? If the requirements were going to be modified, should the bidding process have been reopened to the other bidders and others who might now be able to bid? Should bidding be reopened even if it causes a delay in delivery, increased work for the SSA, etc.?
24. Was it acceptable for Paradyne to submit another manufacturer's system for testing with a Paradyne label on it?
25. Was it acceptable to represent a proposed system as existing, if indeed that is what Paradyne did?
26. Is it ethical for a former SSA employee to take a job negotiating contracts with the SSA for a private company? Did this relationship give Paradyne an unfair advantage over its competition?

3

Understanding Ethical Problems

SECTIONS

OBJECTIVES

After reading this chapter, you will be able to:

- Discuss several ethical theories.
- See how these theories can be applied to engineering situations.

In late 1984, a pressure-relief valve on a tank used to store methyl isocyanate (MIC) at a Union Carbide plant in Bhopal, India accidentally opened. MIC is a poisonous compound used in the manufacture of pesticides. When the valve opened, MIC was released from the tank and a cloud of toxic gas formed over the area surrounding the plant. Unfortunately, this neighborhood was very densely populated. Some two thousand people were killed and thousands more injured as a result of the accident. Many of the injured have remained permanently disabled.

The causes of the accident are not completely clear, but there appear to have been many contributing factors. Pipes in the plant were misconnected, and essential safety systems were either broken or had been taken off-line for maintenance. The effects of the leak were intensified by the presence of so many people living in close proximity to the plant.

Among the many important issues this case brings up are questions of balancing risk to the local community with the economic benefits to the larger community of the state or nation. Undoubtedly, the presence of this chemical plant brought significant local economic benefit. However, the accident at the plant also brought disaster to the local community at an enormous cost in human lives and suffering.

How can we decide if on balance the economic benefit brought by this plant outweighed the potential safety hazards?

In order to answer this question and analyze other engineering ethics cases, we need a framework for analyzing ethical problems. In the previous chapter, we saw how codes of ethics can be used as an aid in analyzing ethical issues. In this chapter, we will examine moral theories and see how they can also be used as a means for analyzing ethical cases such as the Bhopal disaster.

3.1 INTRODUCTION

In this chapter, we will develop moral theories that can be applied to the ethical problems confronted by engineers. Unfortunately, a thorough and in-depth discussion of all possible ethical theories is beyond the scope of this text. Rather, some important theories will be developed in sufficient detail for use in analyzing cases.

Our approach to ethical problem solving will be similar to that taken in other engineering classes. To learn how to build a bridge, you must first learn the basics of physics and apply this physics to engineering statics and dynamics. Only when the basic theory and understanding of these topics has been acquired can problems in structures be solved and bridges built. Similarly, in ethical problem solving, we will need some knowledge of ethical theory to provide a framework for understanding and reaching solutions in ethical problems. In this chapter, we will develop this theoretical framework and apply it to an engineering case. We will begin by looking at the origins of Western ethical thinking.

3.2 A BRIEF HISTORY OF ETHICAL THOUGHT

It is impossible in this text to give a complete history of ethical thinking. Many books, some of them quite lengthy, have already been written on this subject. However, it is instructive to give a brief outline of the origins and development of the ethical principles that will be applied to engineering practice.

The moral and ethical theories that we will be applying in engineering ethics are derived from a Western cultural tradition. In other words, these ideas originated in the Middle East and Europe. Western moral thought has not come down to us from just a single source. Rather, it is derived both from the thinking of the ancient Greeks as well as from ancient religious thinking and writing, starting with Judaism and its foundations.

Although it is easy to think of these two sources as separate, there was a great deal of influence on ancient religious thought by the Greek philosophers. The written sources of the Jewish moral traditions are the Torah and the Old Testament of the Bible and their enumeration of moral laws, including the ten commandments. Greek ethical thought originated with the famous Greek philosophers that are commonly studied in freshman philosophy classes, principally Socrates and Aristotle, who discussed ethics at great length in his *Nichomachean Ethics*. Greek philosophic ideas were melded together with early Christian and Jewish thought and were spread throughout Europe and the Middle East during the height of the Roman Empire.

Ethical ideas were continually refined during the course of history. Many great thinkers have turned their attention to ethics and morals and have tried to provide insight into these issues through their writings. For example, philosophers such as Locke, Kant, and Mill wrote about moral and ethical issues. The thinking of these philosophers is especially important for our study of engineering ethics, since they did not rely on religion to underpin their moral thinking. Rather, they acknowledged that moral principles are universal, regardless of their origin, and are applicable even in secular settings.

Many of the moral principles that we will discuss have also been codified and handed down through the law. So, in discussing engineering ethics, there is a large body of thinking—philosophical, legal, and religious—to draw from. However, even though there are religious and legal origins of many of the moral principles that we will encounter in our study of engineering ethics, it is important to acknowledge that ethical conduct is fundamentally grounded in a concern for other people. It is not just about law or religion.

3.3 ETHICAL THEORIES

In order to develop workable ethical problem-solving techniques, we must first look at several theories of ethics in order to have a framework for decision making. It was mentioned before that ethical problem solving is not as cut and dried as problem solving in most engineering classes. In most engineering classes, there is generally just one theory to consider when tackling a problem. In studying engineering ethics, there are several theories that will be considered. The relatively large number of theories doesn't indicate a weakness in theoretical understanding of ethics or a "fuzziness" of ethical thinking. Rather, it reflects the complexity of ethical problems and the diversity of approaches to ethical problem solving that have been developed over the centuries.

Having multiple theories to apply actually enriches the problem-solving process, allowing problems to be looked at from different angles, since each theory stresses different aspects of a problem. Even though we will use multiple theories to examine ethical problems, each theory applied to a problem will not necessarily lead to a different solution. Frequently, different theories yield the same solution. Our basic ethical problem-solving technique will utilize different theories and approaches to analyze the problem and then try to determine the best solution.

What Is a Moral Theory?

Before looking more closely at individual moral theories, we should start with a definition of what a moral theory is and how it functions. A moral theory defines terms in uniform ways and links ideas and problems together in consistent ways [Harris, Pritchard and Rabins, 1985]. This is exactly how the scientific theories used in other engineering classes function. Scientific theories also organize ideas, define terms, and facilitate problem solving. So, we will use moral theories in exactly the same way that engineering theories are used in other classes.

There are four ethical theories that will be considered here, each differing according to what is held to be the most important moral concept. *Utilitarianism* seeks to produce the most utility, defined as a balance between good and bad consequences of an action, taking into account the consequences for everyone affected. A different approach is provided by *duty ethics*. Duty ethics contends that there are duties that should be performed (for example, the duty to treat others fairly or the duty not to injure others) regardless of whether these acts lead to the most good. *Rights ethics* emphasizes that we all have moral rights, and any action that violates these rights is ethically unacceptable. Like duty ethics, the ultimate overall good of the actions is not taken into account. Finally, *virtue ethics* regards actions as right that manifest good character traits (virtues) and regards actions as bad that display bad character traits (vices); this ethical theory focuses on the type of person we should strive to be.

Utilitarianism

The first of the moral theories that will be considered is utilitarianism. Utilitarianism holds that those actions are good that serve to maximize human well-being. The emphasis in utilitarianism is not on maximizing the well-being of the individual, but rather on maximizing the well-being of society as a whole, and as such it is somewhat of a collectivist approach. An example of this theory that has been played out in this country many times over the past century is the building of dams. Dams often lead to great benefit to society by providing stable supplies of drinking water, flood control, and recreational opportunities. However, these benefits often come at the expense of people who live in areas that will be flooded by the dam and are required to find new homes. Utilitarianism tries to balance the needs of society with the needs of the individual, with an emphasis on what will provide the most benefit to the most people.

Utilitarianism is fundamental to many types of engineering analysis, including risk–benefit analysis and cost–benefit analysis, which we will discuss later. However, as good as the utilitarian principle sounds, there are some problems with it. First, as seen in the example of the building of a dam, sometimes what is best for everyone may be bad for a particular individual or group of individuals. An example of this problem is the proposed Waste Isolation Pilot Plant (WIPP) near Carlsbad, New Mexico. WIPP is designed to be a permanent repository for nuclear waste generated in the United States. It consists of a system of tunnels bored into underground salt formations, which are considered by geologists to be extremely stable, especially to incursion of water, which could lead to seepage of the nuclear wastes into ground water. However, there are many who oppose the opening of this facility, principally on the grounds that transportation of the wastes across highways has the potential for accidents that might cause health problems for people living near these routes.

An analysis of WIPP using utilitarianism might indicate that the disposal of nuclear wastes is a major problem hindering the implementation of many useful technologies, including medicinal uses of radioisotopes and nuclear generation of electricity. Solution of this waste disposal problem will benefit society by providing improved health care and more plentiful electricity. The slight potential for adverse health effects for individuals living near the transportation routes is far outweighed by the overall benefits to society. So, WIPP should be allowed to open. As this example demonstrates, the utilitarian approach can seem to ignore the needs of individuals, especially if these needs seem relatively insignificant.

Another objection to utilitarianism is that its implementation depends greatly on knowing what will lead to the most good. Frequently, it is impossible to know exactly what the consequences of an action are. It is often impossible to do a complete set of experiments to determine all of the potential outcomes, especially when humans are involved as subjects of the experiments. So, maximizing the benefit to society involves guesswork and the risk that the best guess might be wrong. Despite these objections, utilitarianism is a very valuable tool for ethical problem solving, providing one way of looking at engineering ethics cases.

Before ending our discussion of utilitarianism, it should be noted that there are many flavors of the basic tenets of utilitarianism. Two of these are act utilitarianism and rule utilitarianism. Act utilitarianism focuses on individual actions rather than on rules. The best known proponent of act utilitarianism was John Stuart Mill (1806–1873), who felt that most of the common rules of morality (e.g., don't steal, be honest, don't harm others) are good guidelines derived from centuries of human experience. However, Mill felt that individual actions should be judged based on whether the most good was produced in a given situation, and rules should be broken if doing so will lead to the most good.

John Stuart Mill, a leading philosopher of utilitarianism. Courtesy of the Library of Congress.

Rule utilitarianism differs from act utilitarianism in holding that moral rules are most important. As mentioned previously, these rules include "do not harm others" and "do not steal." Rule utilitarians hold that although adhering to these rules might not always maximize good in a particular situation, overall, adhering to moral rules will ultimately lead to the most good. Although these two different types of utilitarianism can lead to slightly different results when applied in specific situations, in this text we will consider these ideas together and not worry about the distinctions between the two.

Cost–Benefit Analysis

One tool often used in engineering analysis, especially when trying to determine whether a project makes sense, is cost–benefit analysis. Fundamentally, this type of analysis is just an application of utilitarianism. In cost–benefit analysis, the costs of project are assessed, as are the benefits. Only those projects with the highest ratio of benefits to costs will be implemented. This principle is similar to the utilitarian goal of maximizing the overall good.

As with utilitarianism, there are pitfalls in the use of cost–benefit analysis. While it is often easy to predict the costs for most projects, the benefits that are derived from them are often harder to predict and to assign a dollar value to. Once dollar amounts for the costs and benefits are determined, calculating a mathematical ratio may seem very objective and therefore may appear to be the best way to make a decision. However, this ratio can't take into account many of the more subjective aspects of a decision. For example, from a pure cost–benefit discussion, it might seem that the building of a dam is an excellent idea. But this analysis won't include other issues such as whether the benefits outweigh the loss of a scenic wilderness area or the loss of an endangered species with no current economic value. Finally, it is also important to determine whether those who stand to reap the benefits are also those who will pay the costs. It is unfair to place all of the costs on one group while another reaps the benefits.

Duty Ethics and Rights Ethics

Two other ethical theories—duty ethics and rights ethics—are similar to each other and will be considered together. These theories hold that those actions are good that respect the rights of the individual. Here, good consequences for society as a whole are not the only moral consideration.

A major proponent of duty ethics was Immanuel Kant (1724–1804), who held that moral duties are fundamental. Ethical actions are those actions that could be written down on a list of duties: be honest, don't cause suffering to other people, be fair to others, etc. These actions are our duties because they express respect for persons, express an unqualified regard for autonomous moral agents, and are universal principles [Martin and Schinzinger, 1989]. Once one's duties are recognized, the ethically correct moral actions are obvious. In this formulation, ethical acts are a result of proper performance of one's duties.

Rights ethics was largely formulated by John Locke (1632–1704), whose statement that humans have the right to life, liberty, and property was paraphrased in the Declaration of Independence of the soon-to-be United States of America in 1776. Rights ethics holds that people have fundamental rights that other people have a duty to respect.

Duty ethics and rights ethics are really just two different sides of the same coin. Both of these theories achieve the same end: Individual persons must be respected, and actions are ethical that maintain this respect for the individual. In duty ethics, people have duties, an important one of which is to protect the rights of others. And in rights ethics, people have fundamental rights that others have duties to protect.

As with utilitarianism, there are problems with the duty and rights ethics theories that must be considered. First the basic rights of one person (or group) may conflict with the basic rights of another group. How do we decide whose rights have priority? Using our previous example of the building of a dam, people have the right to use their property. If their land happens to be in the way of a proposed dam, then rights ethics would hold that this property right is paramount and is sufficient to stop the dam project. A single property holder's objection would require that the project be terminated. However, there is a need for others living in nearby communities to have a reliable water supply and to be safe from continual flooding. Who's rights are paramount here? Rights and duty ethics don't resolve this conflict very well; hence, the utilitarian approach of trying to determine the most good is more useful there.

Immanuel Kant, German philosopher whose work included early formulations of duty ethics. Courtesy of the Library of Congress.

The second problem with duty and rights ethics is that these theories don't always account for the overall good of society very well. Since the emphasis is on the individual, the good of a single individual can be paramount compared to what is good for society as a whole. The WIPP case discussed before illustrates this problem. Certainly, people who live along the route where the radioactive wastes will be transported have the right to live without fear of harm due to accidental spills of hazardous waste. But the nation as a whole will benefit from the safe disposal of these wastes. Rights ethics would come down clearly on the side of the individuals living along the route despite the overall advantage to society.

Already it is clear why we will be considering more than one ethical theory in our discussion of engineering cases. The theories already presented clearly represent different ways of looking at ethical problems and can frequently arrive at different solutions. Thus, any complete analysis of an ethical problem must incorporate multiple theories if valid conclusions are to be drawn.

Virtue Ethics

Another important ethical theory that we will consider is virtue ethics. Fundamentally, virtue ethics is interested in determining what kind of people we should be. Virtue is often defined as moral distinction and goodness. A virtuous person exhibits good and beneficial qualities. In virtue ethics, actions are considered right if they support good character traits (virtues) and wrong if they support bad character traits (vices) [Schinzinger and Martin, 1989]. Virtue ethics focuses on such words as responsibility, honesty, competence, and loyalty, which are virtues; and dishonesty, disloyalty, and irresponsibility, which are vices. As you can see, virtue ethics is closely tied to personal character. We do good things because we are virtuous people and seek to enhance these character traits in ourselves and in others.

In many ways, this theory may seem to be mostly personal ethics and not particularly applicable to engineering or business ethics. However, personal morality cannot, or at any rate should not, be separated from business morality. If a behavior is virtuous for the individual in her personal life, the behavior is virtuous in her business life as well.

How can virtue ethics be applied to business and engineering situations? This type of ethical theory is somewhat trickier to apply to the types of problems that we will consider, perhaps because virtue ethics seems less concrete and less susceptible to rigorous analysis and because it is harder to describe nonhuman entities such as a corporation or government in terms of virtue. However, we can use virtue ethics in our engineering career by answering questions such as: Is this action honest? Will this action demonstrate loyalty to my community and/or my employer? Have I acted in a responsible fashion? Often, the answer to these questions makes the proper course of action obvious.

Personal vs. Corporate Morality

This is an appropriate place to discuss a tricky issue in engineering ethics: Is there a distinction between the ethics practiced by an individual and that practiced by a corporation? Put another way, can a corporation be a moral agent as an individual can? This is a question that is central to many discussions of business and engineering ethics. If a corporation has no moral agency, then it cannot be held accountable for its actions, although sometimes individuals within a company can be held accountable. The law is not always clear on the answer to this question and can't be relied upon to resolve the issue.

This dilemma comes most sharply into focus in a discussion of virtue ethics. Can a company truly be expected to display honesty or loyalty? These are strictly human traits and cannot be ascribed to a corporation. In the strictest definition of moral agency, a company cannot be a moral agent, and yet companies have many dealings with individuals or groups of people.

How, then, do we resolve this problem? In their capacity to deal with individuals, corporations should be considered pseudomoral agents and should be held accountable in the same way that individuals are, even if the ability to do this within the legal system is limited. In other words, with regard to an ethical problem, responsibility for corporate wrongdoing shouldn't be hidden behind a corporate mask. Just because it isn't really a moral agent like a person doesn't mean that a corporation can do whatever it pleases. Instead, in its interactions with individuals or communities, a corporation must respect the rights of individuals and should exhibit the same virtues that we expect of individuals.

Which Theory to Use?

Now that we have discussed four different ethical theories, the question arises: How do we decide which theory is applicable to a given problem? The good news is that in solving ethical problems, we don't have to choose from among these theories. Rather, we can use all of them to analyze a problem from different angles and see what result each of the theories gives us. This allows us to examine a problem from different perspectives to see what conclusion each one reaches. Frequently, the result will be the same even though the theories are very different.

Take, for example, a chemical plant near a small city that discharges a hazardous waste into the groundwater. If the city takes its water from wells, the water supply for the city will be compromised and significant health problems for the community may result. Rights ethics indicates that this pollution is unethical, since it causes harm to many of the residents. A utilitarian analysis would probably also come to the same conclusion, since the economic benefits of the plant would almost certainly be outweighed by the negative effects of the pollution and the costs required to ensure a safe municipal water supply. Virtue ethics would say that discharging wastes into groundwater is irresponsible and harmful to individuals and so shouldn't be done. In this case, all of the ethical theories lead to the same conclusion.

What happens when the different theories seem to give different answers? This scenario can be illustrated by the discussion of WIPP presented previously. Rights ethics indicated that transporting wastes through communities is not a good idea, whereas utilitarianism concluded that WIPP would be beneficial to society as a whole. This is a trickier situation, and the answers given by each of the theories must be examined in detail, compared with each other, and carefully weighed. Generally, rights and duty ethics should take precedence over utilitarian considerations. This is because the rights of individuals should receive relatively stronger weight than the needs of society as a whole. For example, an action that led to the death of even one person is generally viewed very negatively, regardless of the overall benefit to society. After thorough analysis using all of the theories, a balanced judgment can be formed. Some further techniques for forming this judgment will be discussed in Chapter 4.

Non-Western Ethical Thinking

It is tempting to think that the ethical theories that have been described here are applicable only in business relations within cultures that share our Western ethical traditions: Europe and the Americas. Since the rest of the world has different foundations for its ethical systems, it might seem that what we learn here won't be applicable in our business dealings in, for example, Japan, India, Africa, or Saudi Arabia. However, this thinking is incorrect. Ethics is not geographic or cultural. Indeed, ethical thinking has developed similarly around the world and is not dependent on a Western cultural or religious tradition. Ethical standards are similar worldwide.

For example, ethical principles in Arab countries are grounded in the traditions of their religion, Islam. Islam is one of the three major monotheistic religions, along with Christianity and Judaism. It is surprising to many Westerners that Islam developed in the Middle East, just as Judaism and Christianity did, and shares many prophets and religious concepts with the other two monotheistic religions. The foundations of ethical principles relating to engineering and business in Islamic countries are thus very similar to those in Western countries. Although cultural practices may vary when dealing with the many Islamic nations that stretch from Africa and the Middle East to Southeast Asia, the same ethical principles that apply in Western countries are applicable.

Similarly, ethical principles of Hindus, Buddhists, and practitioners of all the world's major religions are similar. Although the ethical principles in other cultures may be derived in different ways, the results are generally the same, regardless of culture.

Moreover, personal ethics are not determined by geography. Personal and business behavior should be the same regardless of where you happen to be on a given day.

For example, few would find the expression "When in Rome, do as the Romans do" applicable to personal morality. If you believe that being deceptive is wrong, certainly it is no less wrong when you are dealing with a (hypothetical) culture where this behavior is not considered to be bad. Thus, the ethics that we discuss in this book will be applicable regardless of where you are doing business.

APPLICATION: CASES

The Disaster at Bhopal

On the night of December 2, 1984, a leak developed in a storage tank at a Union Carbide chemical plant in Bhopal, India. The tank contained 10,000 gallons of methyl isocyanate (MIC), a highly toxic chemical used in the manufacture of pesticides, such as Sevin. The leak sent a toxic cloud of gas over the surrounding slums of Bhopal, resulting in the death of over 2,000 people, and injuries to over 200,000 more.

The leak was attributed to the accidental pouring of water into the tank. Water reacts very vigorously with MIC, causing heating of the liquid. In Bhopal, the mixing of water with MIC increased the temperature of the liquid in the tank to an estimated 400°F. The high temperature caused the MIC to vaporize, leading to a buildup of high pressure within the tank. When the internal pressure became high enough, a pressure-relief valve popped open, leaking MIC vapors into the air.

The water had probably been introduced into the tank accidentally. A utility station on the site contained two pipes side by side. One pipe carried nitrogen, which was used to pressurize the tank to allow the liquid MIC to be removed. The other pipe contained water. It appears that instead of connecting the nitrogen pipe, someone accidentally connected the water pipe to the MIC tank. The accident was precipitated when an estimated 240 gallons of water were injected into the MIC storage tank.

As with many of the disasters and accidents that we study in this book, there was not just one event that led to the disaster, but rather there were several factors that contributed to this accident. Any one of these factors alone probably wouldn't have led to the accident, but the combination of these factors made the accident almost inevitable and the consequences worse. A major factor in this accident was the curtailment of plant maintenance as part of a cost-cutting effort. The MIC storage tank had a refrigeration unit on it, which should have helped to keep the tank temperatures closer to normal, even with the water added, and might have prevented the vaporization of the liquid. However, this refrigeration unit had stopped working five months before the accident and hadn't yet been repaired.

The tank also was equipped with an alarm that should have alerted plant workers to the dangerous temperatures; this alarm was improperly set, so no warning was given. The plant was equipped with a flare tower. This is a device designed to burn vapors before they enter the atmosphere, and it would have been able to at least reduce, if not eliminate, the amount of MIC reaching the surrounding neighborhood. The flare tower was not functioning at the time of the accident. Finally, a scrubber that was used to neutralize toxic vapors was not activated until the vapor release was already in progress. Some investigators pointed out that the scrubber and flare systems were probably inadequate, even had they been functioning. However, had any of these systems been functioning at the time of the accident, the disaster could have at least been mitigated, if not completely averted. The fact that none of them were operating at the time ensured that once the water had been mistakenly added to the MIC tank, the ensuing reaction would proceed undetected until it was too late to prevent the accident.

It is unclear to whom the ultimate blame for this accident should be laid. The plant designers clearly did their job by anticipating problems that would occur and installing safety systems to prevent or mitigate potential accidents. The management of the plant seems obviously negligent. It is sometimes necessary for some safety features to be taken off-line for repair or maintenance. But to have all of the safety systems inoperative simultaneously is inexcusable. Union Carbide also seems negligent in not preparing a plan for notifying and evacuating the surrounding population in the event of an accident. Such plans are standard in the United States and are often required by local ordinance.

Union Carbide was unable to say that such an accident was unforseeable. Leaky valves in the MIC system had been a problem at the Bhopal plant on at least six occasions before the accident. One of these gas leaks involved a fatality. Moreover, Union Carbide had a plant in Institute, West Virginia that also produced MIC. The experience in West Virginia was similar to that in Bhopal before the accident. There had been a total of 28 leaks of MIC over the previous five years, none leading to any serious problems. An internal Union Carbide memo from three months before the Bhopal accident warned of the potential for a runaway reaction in MIC storage tanks in West Virginia and called into question the adequacy of emergency plans at the plants. The memo concluded that "a real potential for a serious incident exists" [*US News and World Report*, Feb. 4, 1985, p. 12]. Apparently, these warnings had not been transmitted to the plant in India.

Ultimately, some share of the blame must be borne by the Indian government. Unlike in most Western nations, there was very little in the way of safety standards under which U.S. corporations must operate. In fact, third-world countries have often viewed pollution control and safety regulation as too expensive, and attempts by the industrialized nations to enforce Western-style safety and environmental regulations worldwide are regarded as attempts to keep the economies of developing countries backward [*Atlantic Monthly*, March 1987, p. 30]. In addition, the local government had no policy or zoning forbidding squatters and others from living so close to a plant where hazardous compounds are stored and used. The bulk of the blame goes to Union Carbide for failure to adequately train and supervise its Indian employees in the maintenance and safety procedures that are taken for granted in similar plants in the United States.

In the aftermath of the accident, lawsuits totaling over $250 billion were filed on behalf of the victims of the accident. Union Carbide committed itself to ensuring that the victims of the accident were compensated in a timely fashion. Union Carbide also helped set up job training and relocation programs for the victims of the accident. Ultimately, it has been estimated that approximately 10,000 of those injured in the accident will suffer some form of permanent damage [*Atlantic Monthly*, March 1987, p. 30].

The Aberdeen Three

The Aberdeen Three is one of the classic cases often used in engineering ethics classes and texts to illustrate the importance of environmental protection and the safety of workers exposed to hazardous and toxic chemicals. The Aberdeen Proving Ground is a U.S. Army weapons development and test center located on a military base in Maryland with no access by civilian nonemployees. Since World War II, Aberdeen has been used to develop and test chemical weapons. Aberdeen has also been used for the storage and disposal of some of these chemicals.

This case involves three civilian managers at the Pilot Plant at the Proving Grounds: Carl Gepp, manager of the Pilot plant; William Dee, who headed the chemical weapons development team; and Robert Lentz, who was in charge of developing manufacturing processes for the chemical weapons. [Weisskopf, 1989] Between 1983 and 1986, inspections at the Pilot Plant indicated that there were serious safety hazards. These hazards included carcinogenic and flammable substances left in open containers, chemicals that can become lethal when mixed together being stored in the same room, barrels of toxic chemicals that were leaking, and unlabeled containers of chemicals. There was also an external tank used to store sulfuric acid that had leaked 200 gallons of acid into a local river. This incident triggered state and federal safety investigations that revealed inadequate chemical retaining dikes and a system for containing and treating chemical hazards that was corroded and leaking.

In June of 1988, the three engineer/managers were indicted for violation of RCRA, the Resource Conservation and Recovery Act. RCRA had been passed by Congress in 1976 and was intended to provide incentives for the recovery of important resources from wastes, the conservation of resources, and the control of the disposal of hazardous wastes. RCRA banned the dumping of solid hazardous wastes and included criminal penalties for violations of hazardous-waste disposal guidelines. The three managers claimed that they were not aware that the plant's storage practices were illegal and that they did things according to accepted practices at the Pilot Plant. Interestingly, since this was a criminal prosecution, the Army could not help defray the costs of the manager's defense, and each of them incurred great costs defending themselves.

In 1989, the three engineer/managers were tried and convicted of illegally storing, treating, and disposing of hazardous wastes. There was no indication that these three were the ones who actually handled chemicals in an unsafe manner, but as managers of the plant, the three were ultimately responsible for how the chemicals were stored and for the maintenance of the safety equipment. The potential penalty for these crimes was up to 15 years in prison and a fine of up to $750,000. Gepp, Dee, and Lentz were each found guilty and sentenced to three years probation and 1,000 hours of community service. The relative leniency of the sentences was based partly on the large court costs each had already incurred.

PROFESSIONAL SUCCESS: TEAMWORK

Ethical issues can arise when working on projects in groups or teams. Many of your engineering classes are designed so that labs or projects are performed in groups. Successful performance in a group setting is a skill that is best learned early in your academic career since most projects in industry involve working as part of a team.

In order for a project to be completed successfully, cooperation among team members is essential. Problems can arise when a team member doesn't do a good job on his part of the project, doesn't make a contribution at all, or doesn't complete his assignments on time. There can also be a problem when one team member tries to do everything. This shuts out teammates who want to contribute and learn. An analogy can be made here to team sports: clearly one individual on the team who is not performing his role can lead to a loss for the entire team. Equally true, individuals who try to do it all—"ballhogs"—can harm the team. Ethical teamwork includes performing the part of the work that you are assigned, keeping to schedules, sharing information with other team members, and helping to foster a cooperative and supportive team atmosphere so everyone can contribute.

KEY TERMS

Utilitarianism
Rights ethics
Duty ethics
Virtue ethics
Cost–benefit analysis

REFERENCES

CHARLES E. HARRIS, JR., MICHAEL S. PRITCHARD, AND MICHAEL J. RABINS, *Engineering Ethics, Concepts and Cases*, Wadsworth Publishing Company, Belmont, CA, 1995.

MIKE W. MARTIN AND ROLAND SCHINZINGER, *Ethics in Engineering*, 2d. ed., McGraw-Hill, New York, 1989.

Bhopal

PHILIP ELMER-DEWITT, "What Happened at Bhopal?" *Time Magazine*, April 1, 1985, p. 71.

"Bhopal Disaster—New Clues Emerge," *US News and World Report*, Feb. 4, 1985, p. 12.

PETER STOLER, "Frightening Findings at Bhopal," *Time Magazine*, Feb. 18, 1985, p. 78.

FERGUS M. BORDEWICH, "The Lessons of Bhopal," *Atlantic Monthly*, March 1987, p. 30.

Aberdeen Three

STEVEN WEISSKOPF, "The Aberdeen Mess," *The Washington Post Magazine*, Jan. 15, 1989, p. 55.

Problems

1. Let's revisit the space shuttle *Challenger* case and analyze it using the ethical theories developed in this chapter. What does utilitarianism tell us about this case? In your analysis, be sure to include issues regarding benefits to the United States and mankind that might result from the space shuttle program. You might also include benefits to Morton Thiokol and the communities where it operates if the program is successful.
2. What do duty and rights ethics tell us about the *Challenger* case? How do your answers to this question and to the previous question influence your ideas on whether the *Challenger* should have been launched?

3. Analyze the Pentium chip, the DIA, and the Paradyne computer cases from Chapter 2 using virtue ethics. Start by deciding what virtues are important for people in these businesses (e.g., honesty, fairness, etc.). Then see if these virtues were exhibited by the engineers working for these companies.

Bhopal

4. Use the ethical theories discussed in this chapter to analyze the Bhopal case. Topics to be considered should include the placing of a hazardous plant in a populated area, decisions to defer maintenance on essential safety systems, etc. Especially important theories will be rights and duty ethics and utilitarianism.
5. Use the code of ethics of the American Institute of Chemical Engineers, reprinted in Appendix A, to analyze what a process engineer working at this plant should have done. What does the code say about the responsibilities of the engineers who designed the plant and the engineers responsible for making maintenance decisions?
6. What responsibility does Union Carbide have for the actions of its subsidiaries? Union Carbide India was 50.9% owned by the parent company.
7. What duty did Union Carbide have to inform local officials in India of the potential dangers of manufacturing and storing MIC in India?
8. Some of Union Carbide's reports hinted strongly that part of the fault lay with the inadequate workforce available in a third-world country such as India. How valid is this statement? What are the ethical implications for Union Carbide if this statement is true?
9. What responsibility should the national and local government in Bhopal have for ensuring that the plant is operated safely?
10. What relative importance should be placed on keeping safety systems operating as compared to maintaining other operations? (Note: From the reports on this accident, there is no indication that Union Carbide skimped on safety to keep production going. Rather, this is a hypothetical question.)
11. In the absence of environmental or safety laws in the locality where it operates, what responsibility does a U.S. corporation have when operating overseas? Does the answer change if the locality does have laws, but they are less strict than ours? What about the ethics of a U.S. corporation selling products overseas that are banned in the United States, such as DDT?

The Aberdeen Three

12. What does utilitarianism tell us about the behavior of the Aberdeen Three? What do duty and rights ethics tell us? In analyzing this, start by determining who is harmed or potentially harmed by these activities and who benefits or potentially benefits from them.
13. Can the actions of these engineer/managers be classified as engineering decisions, management decisions, or both? Ethically, does it matter whether these decisions were engineering or management decisions?
14. Do you think that the Aberdeen Three knew about RCRA? If not, should they have? Does it really matter if they knew about RCRA or not?
15. Do you think that the Aberdeen Three were knowledgeable about the effects of these chemicals and proper storage methods? Should they have been?
16. Were the actions of the Aberdeen Three malicious?
17. In the course of this case, it came out that cleaning up the chemical storage at Aberdeen would have been paid for out of separate Army funds and would not have come from the budgets of the three managers. What bearing does this information have on the case?
18. What should the Aberdeen Three have done differently? Should the lower level workers at the plant have done anything to solve this problem?
19. The bosses of the Aberdeen Three claimed to have no idea about the conditions at the Pilot Plant. Should they have done anything differently? Should they have been prosecuted as well?
20. Apply the code of ethics of one of the professional societies given in Appendix A to this situation. Were the managers guilty of ethical violations according to the code?

4

Ethical Problem-Solving Techniques

In the early 1990s, newspapers began to report on studies indicating that living near electrical-power distribution systems leads to an increased risk of cancer, especially in children. The risk was attributed to the effects of the weak, low-frequency magnetic fields present near such systems. Further reports indicated that there might also be some risk associated with the use of common household items such as electric blankets and clock radios. Predictably, there was much concern among the public about this problem, and many studies were performed to verify these results. Power companies began to look into methods for reducing the fields, and many engineers sought ways to design products that emitted reduced amounts of this radiation.

In designing products and processes, engineers frequently encounter scenarios like the one just described. Nearly everything an engineer designs has some health or safety risk associated with it. Often, as with the case of the weak magnetic fields, the exact nature of the hazard is only poorly understood. How then does an engineer decide whether it is ethical to work on a particular product or process? What tools are there for an engineer who needs to decide which is the ethically correct path to take?

SECTIONS

OBJECTIVES

After reading this chapter, you will be able to:

- Apply these methods to hypothetical and real cases.
- See how flow charting can be used to solve ethical problems.
- Learn what bribery is and how to avoid it.

In this chapter, we will develop analysis and problem-solving strategies to help answer these questions. These techniques will allow us to put ethical problems in the proper perspective and will point us in the direction of the correct solution.

4.1 INTRODUCTION

Now that we have discussed codes of ethics and moral theories, we are ready to tackle the problem of how to analyze and resolve ethical dilemmas when they occur. In solving engineering problems, it is always tempting to look for an appropriate formula, plug in the numbers, and calculate an answer. This type of problem-solving approach, while sometimes useful for engineering analysis problems, is useless for ethical problem solving. As we have seen in the previous chapters, there are theories that help us to frame our understanding of the problem, but there are no formulas and no easy "plug and chug" methods for reaching a solution.

In this chapter, we will examine methods for analyzing ethical problems and see how to apply them. Obviously, some problems are easily solved. If you are tempted to embezzle money from your employer, it is clear that this action is stealing and is not morally acceptable. However, as mentioned previously, many of the situations encountered by practicing engineers are ambiguous or unclear, involving conflicting moral principles. This is the type of problem for which we will most need analysis and problem-solving methods.

4.2 ANALYSIS OF ISSUES IN ETHICAL PROBLEMS

A first step in solving any ethical problem is to completely understand all of the issues involved. Once these issues are determined, frequently a solution to the problem becomes apparent. The issues involved in understanding ethical problems can be split into three categories: factual, conceptual, and moral [Harris, Pritchard, and Rabins, 1985]. Understanding these issues helps to put an ethical problem in the proper framework and often helps point the way to a solution.

Types of Issues in Ethical Problem Solving

Let's begin by examining in depth each of the types of issues involved in ethical problems. Factual issues involve what is actually known about a case—i.e., what the facts are. Although this concept seems straightforward, the facts of a particular case are not always clear and may be controversial. An example of facts that are not necessarily clear can be found in the controversy in contemporary society regarding abortion rights. There is great disagreement over the point at which life begins and at which point a fetus can be legally protected. *Roe v. Wade*, the original decision legalizing abortion in the United States, was decided by the Supreme Court by a split decision. Even the justices of the Supreme Court were unable to agree on this "fact."

In engineering, there are controversies over facts as well. For example, global warming is of great concern to society as we continue to emit greenhouse gases into the atmosphere. Greenhouse gases, such as carbon dioxide, trap heat in the atmosphere. This is thought to lead to a generalized warming of the atmosphere as emissions from automobiles and industrial plants increase the carbon dioxide concentration in the atmosphere. This issue is of great importance to engineers, since they might be required to design new products or redesign old ones to comply with stricter environmental standards if this warming effect indeed proves to be a problem. However, the global warming process is only barely understood, and the need to curtail emission of these gases is a controversial topic. If it were known exactly what the effects of emitting greenhouse gases into the atmosphere would be, the engineer's role in reducing this problem would be clearer.

Conceptual issues have to do with the meaning or applicability of an idea. In engineering ethics, this might mean defining what constitutes a bribe as opposed to an acceptable gift, or determining whether certain business information is proprietary. In the case of the bribe, the value of the gift is probably a well-known fact. What isn't known is whether accepting it will lead to unfair influence on a business decision. For example, conceptually it must be determined if the gift of tickets to a sporting event by a potential supplier of parts for your project is meant to influence your decision or is just a nice gesture between friends. Of course, like factual issues, conceptual issues are not always clear-cut and will often result in controversy as well.

Once the factual and conceptual issues have been resolved, at least to the extent possible, all that remains is to determine which moral principle is applicable to the situation. Resolution of moral issues is often more obvious. Once the problem is defined, it is usually clear which moral concept applies, and the correct decision becomes obvious. In our example of a "gift" offered by a sales representative, once it is determined whether it is simply a gift or is really a bribe, then the appropriate action is obvious. If we determine that it is indeed a bribe, then it cannot ethically be accepted.

Given that the issues surrounding an ethical problem can be controversial, how can these controversies be resolved? Factual issues can often be resolved through research to establish the truth. It is not always possible to achieve a final determination of the "truth" that everyone can agree on, but generally, further research helps clarify the situation, can increase the areas of agreement, and can sometimes achieve consensus on the facts. Conceptual issues are resolved by agreeing on the meaning of terms and concepts. Sometimes agreement isn't possible, but as with factual issues, further analysis of the concepts at least clarifies some of the issues and helps to facilitate agreement. Finally, moral issues are resolved by agreement as to which moral principles are pertinent and how they should be applied.

Often, all that is required to solve a particular ethical problem is a deeper analysis of the issues involved according to the principles given previously. Once the issues are analyzed and agreement is reached on the applicable moral principles, it is clear what the resolution should be.

Application to a Previous Case: Paradyne Computers

To illustrate the use of this problem-solving method, let's revisit the Paradyne computer case presented in Chapter 2, looking at the factual issues first. Clearly, the request for proposals specified that only existing systems would be considered. Paradyne did not have any such system running and had never tested the operating system on the product they actually proposed to sell to the Social Security Administration (SSA). The employment of a former SSA worker by Paradyne to help lobby SSA for the contract is also clear. In this case, the factual issues do not appear particularly controversial.

The conceptual issues involve whether bidding to provide an off-the-shelf product when the actual product is only in the planning stages is lying or is an acceptable business practice. Is placing a Paradyne label over the real manufacturer's label deceptive? Does lobbying your former employer on behalf of your current employer constitute a conflict of interest? These questions will certainly generate discussion. Indeed, Paradyne asserted that it had done nothing wrong and was simply engaging in common business practices. The issue of the conflict of interest is so hard to decide that laws have been enacted making it illegal for workers who have left government employ to lobby their former employers for specified periods of time.

The moral issues then include: Is lying an acceptable business practice? Is it alright to be deceptive if doing so allows your company to get a contract? The answers to these questions are obvious: Lying and deceit are no more acceptable in your business life than in your personal life. So, if conceptually we decide that Paradyne's practices were deceptive, then our analysis indicates that their actions were unethical.

4.3 LINE DRAWING

The line-drawing technique that will be described in this section is especially useful for situations in which the applicable moral principles are clear, but there seems to be a great deal of "gray area" about which ethical principle applies. Line drawing is performed by drawing a line along which various examples and hypothetical situations are placed. At one end is placed the "positive paradigm," an example of something that is unambiguously morally acceptable. At the other end, the "negative paradigm," an example of something that is unambiguously not morally acceptable, is placed. In between is placed the problem under consideration, along with other similar examples. Those examples that more closely conform to the positive paradigm are placed near it, and examples closer to the negative paradigm are placed near that paradigm. By carefully examining this continuum and placing the moral problem under consideration in the appropriate place along the line, it is possible to determine whether the problem is more like the positive or negative paradigm and therefore whether it is acceptable or unacceptable.

Let's illustrate this technique using a hypothetical situation. Our company would like to dispose of a slightly toxic waste by dumping it into a local lake from which a nearby town gets its drinking water. How can we determine if this practice is acceptable? Let's start by defining the problem and the positive and negative paradigms.

Problem: It is proposed that our company dispose of a slightly hazardous waste by dumping it into a lake. A nearby town takes its drinking water supply from this lake. Our research shows that with the amount of waste we plan to put into the lake, the average concentration of the waste in the lake will be 5 parts per million (ppm). The EPA limit for this material has been set at 10 ppm. At the 5-ppm level, we expect no health problems, and consumers would not be able to detect the compound in their drinking water.

Positive paradigm: The water supply for the town should be clean and safe.

Negative paradigm: Toxic levels of waste are put into the lake.

Let's start by drawing a line and placing the positive and negative paradigms on it:

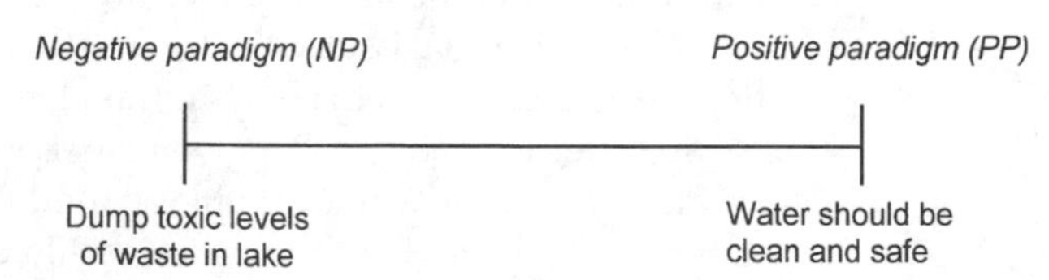

Figure 4.1. Example of line drawing showing the placement of the negative and positive paradigms.

Now let's establish some other hypothetical examples for consideration:

1. The company dumps the chemical into the lake. At 5 ppm, the chemical will be harmless, but the town's water will have an unusual taste.
2. The chemical can be effectively removed by the town's existing water-treatment system.
3. The chemical can be removed by the town with new equipment that will be purchased by the company.
4. The chemical can be removed by the town with new equipment for which the taxpayer will pay.
5. Occasionally, exposure to the chemical can make people feel ill, but this only lasts for an hour and is rare.
6. At 5 ppm, some people can get fairly sick, but the sickness only lasts a week, and there is no long-term harm.
7. Equipment can be installed at the plant to further reduce the waste level to 1 ppm.

Obviously, we could go on for a long time creating more and more test examples. Generally, where your problem fits along the line is obvious with only a few examples, but the exercise should be continued with more examples until it is clear what the proper resolution is. Now let's redraw our line with the examples inserted appropriately:

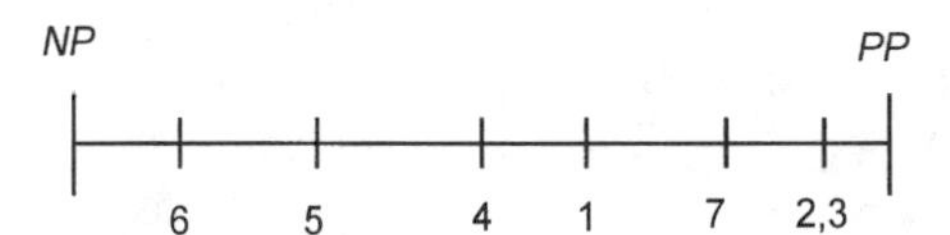

Figure 4.2. Same as Figure 4.1, with the addition of the examples to the line.

After setting up the examples, it may be clear that there is a gap in the knowledge. For example, in our case, we might need more information on seasonal variations in waste concentration and water usage of the town. We also could use information on potential interactions of the chemical with other pollutants, such as runoff of pesticides from local farms. Note that there is some subjectivity in determining exactly where along the line each of the examples fits.

Now let's complete the exercise by denoting our problem by a "P" and inserting it at the appropriate place along the line. As with the previous examples, placement of the problem along the line is somewhat subjective.

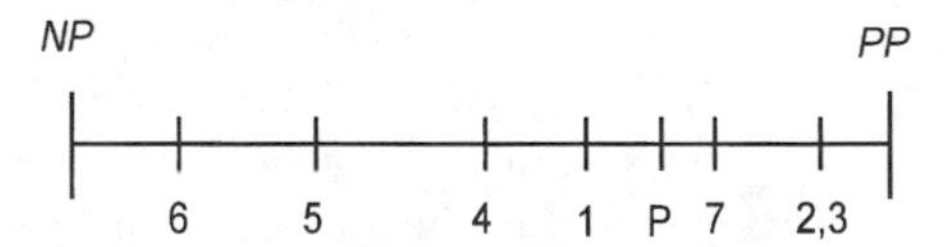

Figure 4.3. Final version of the line-drawing example, with the problem under consideration added.

As drawn here, it is clear that dumping the toxic waste is probably a morally acceptable choice, since no humans will be harmed and the waste levels will be well below those that could cause any harm. However, since it is somewhat far from the positive paradigm, there are probably better choices that can be made, and the company should investigate these alternatives.

It should be noted that although this action seems ethically acceptable, there are many other considerations that might be factored into the final decision. For example, there are political aspects that should also be considered. Many people in the community are likely to regard the dumping of a toxin at any level as unacceptable. Good community relations might dictate that another solution should be pursued instead. The company also might want to avoid the lengthy amount of time required to obtain a permit for the dumping and the oversight by various government units. This example illustrates that line drawing can help solve the ethical aspects of a problem, but a choice that appears morally acceptable still might not be the best choice when politics and community relations are considered as well. Of course, the immoral choice is never the correct choice.

Although this problem-solving method seems to help with problem analysis and can lead to solutions, there are many pitfalls in its use. If not used properly, line drawing can lead to incorrect results. For example, line drawing can easily be used to prove that something is right when it is actually wrong. Line drawing is only effective if it is used objectively and honestly. The choice of where to put the examples and how to define the paradigms is up to you. You can reach false conclusions by using incorrect paradigms, by dishonest placement of the examples along the line, and by dishonest placement of the problem within the examples. In our example, we might have decided that the problem is somewhat like example 2 and thus placed our problem closer to the positive paradigm, making this solution seem more acceptable. Line drawing can be a very powerful analytic tool in ethical problems, but only if used conscientiously.

There is a long history of the improper use of this technique. In its early days, this method was known as "casuistry," a term that eventually came to be pejorative. In the Middle Ages, casuistry was often used in religious debates to reach false conclusions. Indeed, one of the definitions of casuistry from the American Heritage Dictionary implies the use of false and subtle reasoning to achieve incorrect solutions. Because of this negative connotation, the term "casuistry" is rarely used any more. This emphasizes the hazards of using line drawing: It is useful only if properly applied.

Application of Line Drawing to the Pentium Chip Case

Now let's apply line drawing to one of our previous cases. In Chapter 2, there is a discussion of the problems that the Intel corporation had in 1994–95 when it was discovered and widely reported in the press that the latest version of the Pentium computer chip had flaws in it. At first, Intel sought to hide this information, but later came around to a policy of offering consumers chips in which the flaw had been corrected. We can use line drawing to get some insight into this problem.

For our positive paradigm, we will use the statement that "products should perform as advertised." The negative paradigm will be: "Knowingly sell products that are defective and that will negatively affect customer's applications." A few examples that we can add to the line are:

1. There is a flaw in the chip, but it truly is undetectable and won't affect any customer's applications.
2. There are flaws in the chip, the customer is informed of them, but no help is offered.
3. A warning label says that the chip should not be used for certain applications.
4. Recall notices are sent out, and all flawed chips are replaced.
5. Replacement chips are offered only if the customer notices the problem.

Of course, there are many other possible examples. One view of the line, then, is as follows:

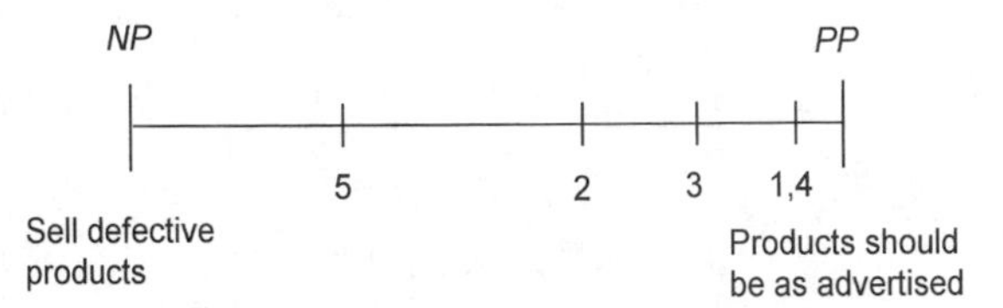

Figure 4.4. Application of line drawing to the Pentium case. Negative and positive paradigms are provided along with the examples.

Where does our situation—there is a flaw, customers aren't informed, and the magnitude of the problem is minimized—fit on this line? One possible analysis is the following:

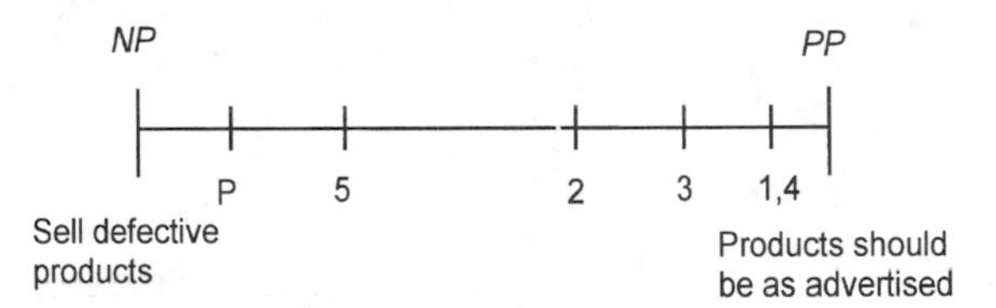

Figure 4.5. Final version of the Pentium chip line-drawing example, with the problem added to the line.

According to this line-drawing analysis, the approach taken by Intel in this case wasn't the best ethical choice.

4.4 FLOW CHARTING

Flow charts are very familiar to engineering students. They are most often used in developing computer programs, and also find application in other engineering disciplines. In engineering ethics, flow charting will be helpful for analyzing a variety of cases, especially those in which there is a sequence of events to be considered or a series of consequences that flows from each decision. An advantage of using a flow chart to analyze ethical problems is that it gives a visual picture of a situation and allows you to readily see the consequences that flow from each decision.

As with the line-drawing technique described in the previous section, there is no unique flow chart that is applicable to a given problem. In fact, different flow charts can be used to emphasize different aspects of the same problem. As with line drawing, it will be essential to be as objective as possible and to approach flow charting honestly. Otherwise, it will be possible to draw any conclusion you want, even one that is clearly wrong.

We can illustrate this technique by applying a simple flow chart to one of the cases that we have previously studied: the disaster at Bhopal. One possible flow chart, illustrated in Figure 4.6, deals with the decision-making process that might have gone on at Union Carbide as they decided whether or not to build a plant at Bhopal. This chart emphasizes safety issues for the surrounding community. As indicated on the chart, there were several paths that might have been taken and multiple decisions that had to be made. The flow chart helps to visualize the consequences of each decision and indicates both the ethical and unethical choices. Of course, this flow chart should be much larger and more complex to thoroughly cover the entire problem.

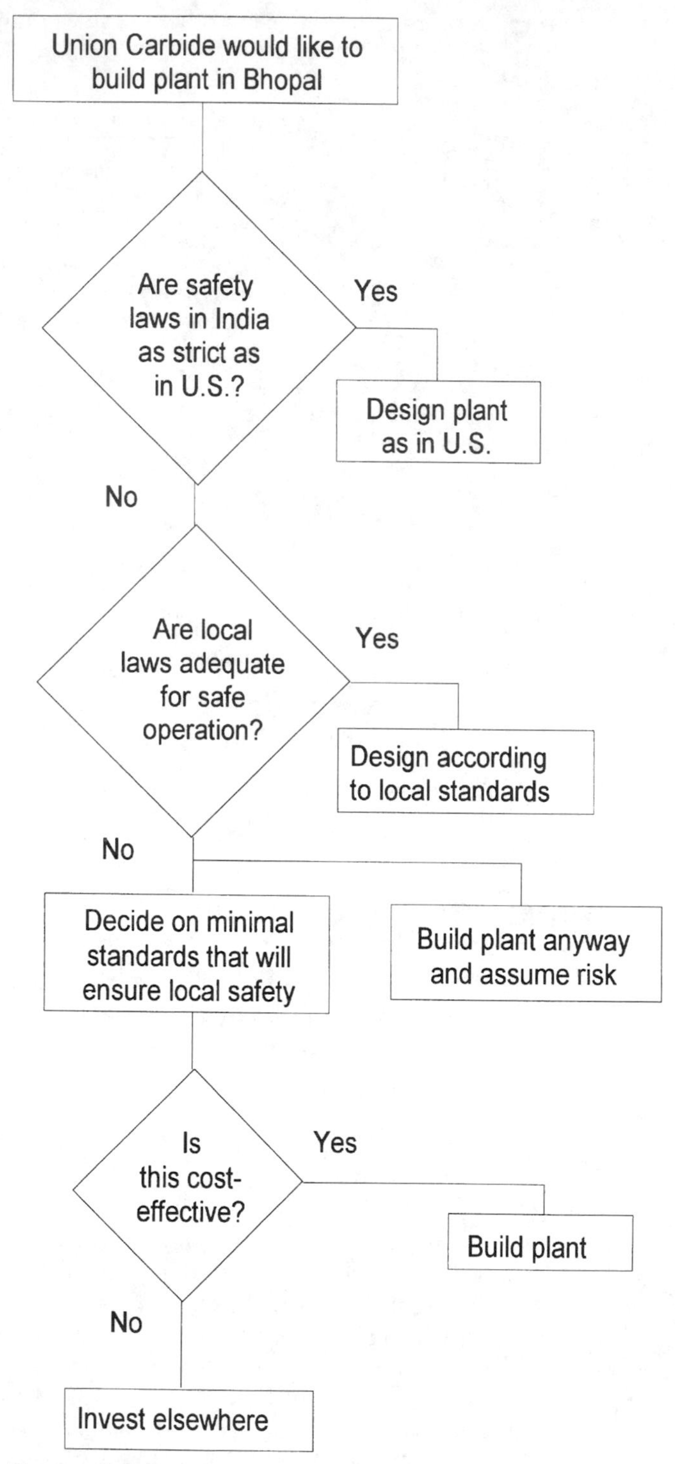

Figure 4.6. Application of a simple flow chart to the Bhopal case, emphasizing potential decisions made during consideration of locating a plant in India.

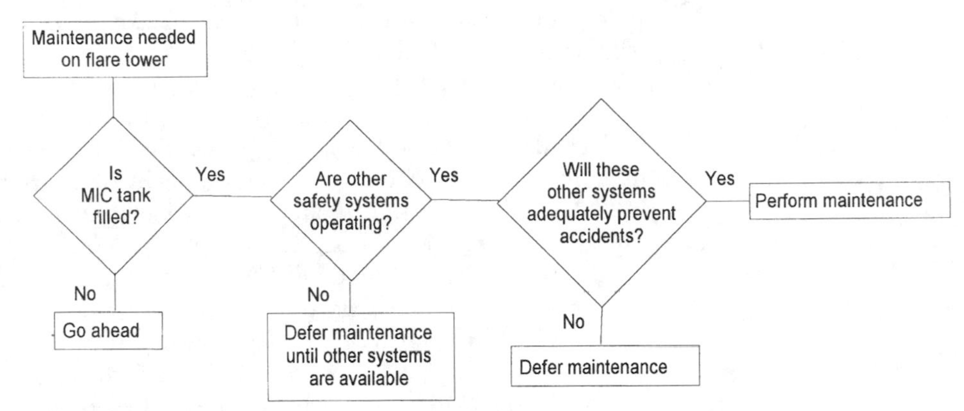

Figure 4.7. An alternative flow chart for the Bhopal case, emphasizing decisions made when considering deactivating the flare tower for maintenance.

Another possible flow chart is shown in Figure 4.7. This chart deals with the decisions required during the maintenance of the flare tower, an essential safety system. It considers issues of whether the MIC tank was filled at the time that the flare tower was taken off-line for maintenance, whether other safety systems were operating when the flare tower was taken out of operation, and whether the remaining safety systems were sufficient to eliminate potential problems. Using such a flow chart, it is possible to decide whether the flare tower can be taken off-line for maintenance or whether it should remain operating.

The key to effective use of flow charts for solving ethical problems is to be creative in determining possible outcomes and scenarios and also to not be shy about getting a negative answer and deciding to stop the project.

4.5 CONFLICT PROBLEMS

An area of ethical problem solving that we will frequently encounter in this book relates to problems that present us with a choice between two conflicting moral values, each of which seems to be correct. How do we make the correct choice in this situation?

Conflict problems can be solved in three ways [Harris, Pritchard, and Rabins, 1995]. Often, there are conflicting moral choices, but one is obviously more significant than the other. For example, protecting the health and safety of the public is more important than your duty to your employer. In this type of case, the resolution of the conflict involves an easy choice.

A second solution is sometimes called the "creative middle way" [Harris, Pritchard, and Rabins, 1995]. This solution is an attempt at some kind of a compromise that will work for everyone. The emphasis here should be on the word "creative," because it takes a great deal of creativity to find a middle ground that is acceptable to everyone and a great deal of diplomacy to sell it to everyone. The sales job is especially difficult because of the nature of compromise, which is often jokingly defined as "the solution where nobody gets what they want." An example of a creative middle ground would be that rather than dumping a toxic waste into a local lake, one finds ways to redesign the production process to minimize the waste product, finds ways to pretreat the waste to minimize the toxicity, or offers to pay for and install the equipment at the municipal water system necessary to treat the water to remove this chemical before it is sent to homes. Obviously, no one will be completely satisfied with these alternatives, since redesigns and pretreatment cost money and take time. Some people will not be satisfied with even a minimized dumping of toxics.

Finally, when there is no easy choice and attempts to find a middle ground are not successful, all that is left is to make the hard choice. Sometimes, you have to bite the bullet and make the best choice possible with the information available at the time. Frequently, you must rely on "gut feelings" for which path is the correct one.

Let's illustrate the resolution-of-conflict problems by revisiting the *Challenger* explosion, which was discussed in Chapter 1, focusing on the dilemma faced by the engineering manager, Bob Lund. The conflict was clear: There was an unknown probability that the shuttle would explode, perhaps killing all aboard. On the other hand, Lund had a responsibility to his company and the people who worked for him. There were consequences of postponing the launch, potentially leading to the loss of future contracts from NASA, the loss of jobs to many Thiokol workers, and perhaps even bankruptcy of the company. For many, the easy choice here is simply to not launch. The risk to the lives of the astronauts is too great and far outweighs any other considerations. It is impossible to balance jobs against lives. After all, most people who lose their jobs will be able to find other employment. However, not everyone will find this to be such an easy choice; clearly, Lund didn't find it to be so.

The creative middle ground might involve delaying the launch until later in the day, when the temperature will have warmed up. Of course, this option might not be possible for many reasons associated with the timing of rocket launches and the successful completion of the planned missions. Instead, perhaps, the astronauts could be informed of the engineer's concerns and be allowed to make the choice whether to launch or not. If a risk is informed and a choice is made by those taking the risk, it somewhat relieves the company of the responsibility if an accident occurs.

The hard choice is what Lund made. He chose to risk the launch, perhaps because the data were ambiguous. He might also have wanted to help ensure the future health of the shuttle program and to save the jobs of the Thiokol workers. As we know, his gamble didn't pay off. The shuttle did explode, causing the deaths of the astronauts and leading to lengthy delays in the shuttle program, political problems for NASA, and business difficulties for Thiokol.

4.6 AN APPLICATION OF PROBLEM-SOLVING METHODS: BRIBERY/ACCEPTANCE OF GIFTS

One of the many gray areas of engineering ethics is the acceptance of gifts from vendors or the offering of gifts to customers to secure business. The difficulty here comes because of the potential for gifts to become bribes or to be perceived of as bribes. Frequently, engineers find themselves in the position of either dealing with vendors who wish to sell them products for incorporation into the engineer's work or acting as vendors themselves and working on sales to other engineers or companies. In this section, we will look at what bribery is and see how some of the problem-solving techniques developed in this chapter can be used to decide when a gift is really a bribe.

Bribery is illegal in the United States and, contrary to popular opinion, is also illegal everywhere in the world. There are some places where bribery may be overlooked, or even expected, but it always takes place "under the table" and is never a legitimate business practice. Moreover, United States federal law forbids American businesses from engaging in bribery overseas, regardless of the local customs or expectations. In many cases, there is a fine line between bribery and a simple gift. Sometimes, the distinction has to do with the value of the gift. Always, it has to do with the intent of the gift. It is important to ensure that no matter how innocent the gift may be, the appearance of impropriety is avoided.

By definition, a bribe is something, such as money or a favor, offered or given to someone in a position of trust in order to induce him to act dishonestly. It is something offered or serving to influence or persuade. What are the ethical reasons for not tolerating bribery? First, bribery corrupts our free-market economic system and is anticompetitive. Unlike the practice of buying the best product at the best price, bribery does not reward the most efficient producer. One can argue the virtues or vices of the free-market economy, but it is the system under which our economy operates, and anything that subverts this system is unfair and unethical. Second, bribery is a sellout to the rich. Bribery corrupts justice and public policy by allowing rich people to make all the rules. In business, it guarantees that only large, powerful corporations will survive, since they are more capable of providing bribes. A small start-up company doesn't have the resources to compete in an environment where expensive favors are required to secure business. Finally, bribery treats people as commodities that can be bought and sold. This practice is degrading to us as human beings and corrupts both the buyer and the seller. [Harris, Pritchard, and Rabins, 1995]

When Is a Gift a Bribe?

Frequently, the boundary between a legitimate gift and a bribe is very subtle. Gifts of nominal value, such as coffee mugs or calendars with a vendor's logo and phone number on it, are really just an advertising tool. Generally, there is no problem with accepting these types of items. Dining with a customer or a supplier is also an acceptable practice, especially if everyone pays his or her own way. It is important from the point of view of both suppliers and customers that good relations be maintained so that good service can be provided. Social interaction, such as eating together, often facilitates the type of close and successful interactions required by both sides. However, when meals or gifts are no longer of low cost and the expense of these items are not shared equally, the possibility for abuse becomes large.

Examples of Gifts vs. Bribes

To help illustrate the difference between bribes and legitimate gifts, let's look at a few potential scenarios to see how fuzzy this boundary can be. No answer will be given to the questions posed, but rather the solution of these questions will be left to the reader.

- During a sales visit, a sales representative offers you a coffee mug with his company's name and logo on it. The value of the mug is five dollars. Can you accept this item? Does the answer to this question change if this item is a $350 crystal bowl with the name of the company engraved on it? How about if there is no engraving on it?
- Your meeting with a sales representative is running into the lunch hour. She invites you to go out for lunch. You go to a fast-food restaurant and pay for your own lunch. Is this practice acceptable? Does the answer to this question change if you go to an expensive French restaurant? If she pays for lunch?
- A sales representative from whom you often purchase asks if you would like to play tennis with him this weekend at one of the local municipal courts. Should you go? Is the answer to this question different if the match is at an exclusive local club to which he belongs? What if he pays the club's guest fee for you?
- A company sales representative would like for you to attend a one-day sales seminar in Cleveland. Your company will pay for your trip. Should you go? How about if the meeting is in Maui? What if the sales representative's company is going to pay for you to go? What if your family is invited as well?

Do the answers to any of these questions change if the gift is offered before you purchase anything from the company, as opposed to after you are already a steady customer? (A more detailed version of these types of scenarios can be found in [Harris, Pritchard, and Rabins, 1995].)

Keep in mind that gifts accepted even after the purchase of something from a company might be a bribe directed at securing future sales from you or might be aimed at engineers at other companies. Although nothing was said about a gift up front, now that you have received one, the expectation of gifts might affect your future purchase decisions. Similarly, an employee of a company like yours might become aware of the gift that you received. He now realizes that if he orders parts from the same supplier that you did, he will receive a gift similar to yours. He will be tempted to order from this supplier even if there is a better supplier of that product on the market. These types of gifts tend to shut out smaller companies that can't necessarily afford gifts and might also cause an increase in everyone's costs, since if everyone now expects to receive gifts, the product cost must go up. Clearly, bribery is pernicious, and even the appearance of bribery should be avoided.

Problem Solving

How can the analysis methods described in this chapter be applied to these examples concerning bribery and the acceptance of gifts? We won't go into the answer to this question in depth here, but will rather save it for the questions at the end of the chapter. However, some general ideas can be presented now. Bribery can easily be analyzed by looking at the factual, conceptual, and moral issues described previously. Frequently, the facts will be obvious: who offered a gift, what its value was, and what its purpose was. Conceptual issues will be somewhat more difficult, since it must be determined whether the gift is of sufficient value to influence a decision or whether that influence is the intent of the gift. Once the conceptual issues have been worked out and it is clear whether or not the gift is a bribe, the moral issue is often very clear.

Line drawing can be very effectively applied to the examples given previously. The subtle differences between the value of the gift, the timing of the gift, etc. are easily visualized using line drawing, and often it will be very clear what the ethical choice will be based on a well-drawn line. Likewise, flow charting can be used to examine the consequences that will result from the acceptance or offer of a gift.

Avoiding Bribery Problems

How does one ensure that accepting a gift doesn't cross the line into bribery? The first and most important method for determining this is to look at company policy. All large corporations and many smaller companies have very clear rules about what is acceptable. Some companies have very strict policies. For example, some companies say that employees are not allowed to accept anything from a vendor and that any social interaction with vendors or customers must be paid for by your company. Any deviation from this rule requires approval from appropriate supervisors. This philosophy is rooted in a sense of trying to avoid any conflict of interest and any appearance of impropriety.

Other companies realize the importance of social interactions in business transactions and allow their employees more discretion in determining what is proper. In the absence of strict corporate guidelines, a preapproval from one's management is the best guide to what is acceptable.

In the absence of any corporate guidelines, another method for determining the acceptability of an action is sometimes referred to as the "New York Times Test": Could your actions withstand the scrutiny of a newspaper reporter? Could you stand to see your name in the newspaper in an article about the gift you received? If you couldn't easily defend your action without resorting to self-serving rationalizations, then you probably shouldn't do it.

APPLICATION: CASES

Low-Frequency Electromagnetic Fields

This case will seem different from many of the other cases that we will study, since there is no disaster or wrongdoing that has to be analyzed after the fact. Rather, this is a case about the experimental nature of engineering, and deals with issues of what engineers should do early in the design cycle for a new product or system to avoid potential harm to customers or the public in general.

By 1994, there had been several studies which suggested that there was a link between weak low-frequency magnetic fields and cancer or other health problems. The effects of low-level radiation, especially from electrical-power distribution systems, first received widespread attention as the result of studies of leukemia occurrences in residential areas of Denver. These studies, by Wertheimer and Leeper, indicated that the incidence of childhood leukemia was correlated to the proximity of the child's home to transformers for residential electrical distribution. Although the correlation found in this study was small, it was "statistically significant." This study received widespread media attention.

Subsequently, there were many other studies both in the United States and Europe that tried to verify these findings, including studies of workers exposed to radiation from cathode-ray tubes (such as computer monitors), workers for electric utilities, etc. The supposed dangers of exposure to electromagnetic radiation were sensationalized by Paul Broduer in several articles in the *New Yorker* magazine that were later published in books, including *Currents of Death* and *The Great Power Line Coverup*.

We should look a little more closely at the types of studies that were performed. Typically, these studies were epidemiological studies and were retrospective looks at populations that might have been exposed to such fields. The studies relied heavily on death certificates to verify the cause of death and various records to determine where the subject lived. One problem with this application of epidemiological methodology is that the exact level of radiation to which the subject was exposed is difficult to determine, especially in retrospect. The results of these studies were controversial, and not all research led to the same conclusion. In fact, as more refined and controlled studies were performed, the harmful effects of the fields seemed to diminish.

Laboratory studies were also performed to determine the biological effects of magnetic fields. These studies were typically performed on cell cultures or on laboratory rodents, but were never performed on humans. (Such testing on humans in the medical field is a ripe area for discussions of ethics!) The results of these studies were conflicting and inconclusive. Moreover, since these studies hadn't been performed on humans, the relationships of any results to human health were debatable.

What was an engineer in the early 90s to do when designing a product that emitted magnetic fields? Keep in mind that a wide variety of common household items have been found to emit magnetic fields, including toasters, electric blankets, and even the clock radio sitting at many bedsides. Some products can be redesigned to reduce or eliminate this problem, but, of course, any design that will lead to reduced emissions will probably cost more. What is the prudent and ethical thing to do when designing such products in an atmosphere where some doubt about safety exists? This case illustrates the problems that engineers have in dealing with and managing the unknown. Many of the designs that engineers produce are experimental in nature or deal with effects that aren't fully understood. It is incumbent on the designer to ensure that she is informed about the potential risks to users of her designs and that she seeks to minimize these risks to the extent possible.

More recently, the evidence for health effects of these fields has been reviewed by panels of several professional societies. Both the IEEE and the American Physical Society (APS) have concluded that there is no evidence indicating that there are any harmful effects, although critics suggest that both of these organizations have vested interests in obtaining this finding. It seems that, for now, the concern over low-frequency electromagnetic radiation is unfounded. The following question still remains: What should the prudent engineer do today when designing products that will emit this type of radiation? And more generally, what should any engineer do when designing products with minimal and/or unknown hazards?

Vice President Spiro Agnew and Construction Kickbacks in Maryland

In January of 1973, architects and consulting engineers all over Baltimore, Maryland were seeking out any available attorneys with experience in criminal law. This activity was brought on by subpoenas issued by the U.S. Attorney for Maryland, George Beall, who was looking into charges of bribes and kickbacks given to elected officials by engineers working in the construction industry. The subpoenas required these engineers to submit the records of their firms to the U.S. attorney. One of these engineers was Lester Matz, a partner in Matz, Childs and Associates, a Baltimore engineering firm. The subsequent events described by Richard Cohen and Jules Witcover in their book *A Heartbeat Away* eventually led to the disgrace and resignation of Spiro Agnew, then the Vice President of the United States.

Matz was an engineer trained at Johns Hopkins University in Baltimore. Although his firm was doing well, it always seemed to lose out to other firms on big public-works contracts. In Maryland, engineering and architectural services for government projects were not put out for bid, but rather were awarded to individual firms using various criteria, including the firm's ability to do the work, its performance on past contracts, etc. Interestingly, unlike the situation for engineering services, the contractor for government projects was chosen through a competitive bidding process. It became clear to Matz that in acquiring government contracts, his talents and those of his firm were unimportant. What was required to get the contracts for public works was contacts in government and the requisite bribes and kickbacks.

In 1961, Matz began courting Spiro T. Agnew, an ambitious and rising politician. In 1962, Matz donated $500 to Agnew's campaign for Baltimore county executive, a post that is roughly equivalent to mayor for the areas of the county outside the city limits of Baltimore. The county executive wielded great power in determining who received contracts for the engineering services required for the numerous public-works projects undertaken by the county. The campaign contribution was given by Matz and his partner in the hopes of receiving some of the county engineering contracts that they had been locked out of. After Agnew won the election, the contribution made by Matz's engineering firm was rewarded with contracts for county engineering work. In return, the firm paid Agnew 5% of their fees from the county work, which apparently was the kickback paid by other engineering firms at the time.

With this arrangement, Matz, Childs and Associates prospered and Matz became relatively wealthy. At its peak, the firm employed nearly 350 people. Matz was able to rent an apartment in Aspen for his winter ski vacations and also had a beach condo at St. Croix in the Virgin Islands. Matz's St. Croix condo was near a condo owned by his friend, Spiro Agnew. The "business" arrangement between Agnew and Matz continued when Agnew was elected governor of Maryland, only now Matz, Childs and Associates received contracts for state work. The financial arrangement remained the same: Agnew received a payment for every contract awarded.

These payments continued even after Agnew was elected vice president on the Republican ticket with Richard Nixon in 1968. Matz testified that he met with Agnew in his office in the White House and had given him an envelope containing $10,000 in cash. Indeed, Matz also indicated that he had given $2,500 dollars to Agnew for a federal contract that a subsidiary of Matz, Child and Associates had received. All told, Matz described payments that he had given Agnew over the years totaling over $100,000.

As a brief aside, it is interesting to describe how the money paid to Agnew was generated. Clearly, these payments had to be made in cash in order to avoid leaving records of the transactions. However, engineering firms are not paid in cash for their services and thus don't typically have large amounts of cash on hand. One method of generating cash was to give cash "bonuses" to key employees. After retaining a sufficient amount to pay the income taxes on the bonus, the employee returned the cash to the firm, where it was placed in a safe until needed. Of course, this practice is a violation of the tax code: The company books record the transaction as a bonus, yet much of the money is retained by the firm. This practice subjected Matz, Child and Associates to prosecution under the federal tax code. This method didn't always generate the required amount of cash, so other means were also used. For example, large "loans" were made to colleagues, who cashed the money and returned it to the firm. These loans were then "repaid" slowly over a long period of time to make the books appear right.

With federal prosecutors threatening to indict Matz and Childs for income-tax evasion and other charges, they decided to provide evidence to the government of the wrongdoing of Agnew and his successor as county executive. Agnew's lawyers and the prosecutors reached an agreement whereby Agnew would resign as vice president and plead *nolo contendere* (no contest) to a single count of income-tax evasion, a felony, for payments received in 1967. This plea is the legal equivalent of a plea of guilty; the defendant doesn't admit to the crime, but does acknowledge that there is enough evidence to convict him. On October 10, 1973, Agnew resigned as vice president, the first vice president to have resigned in disgrace. Later that day, in a dramatic appearance in a Maryland courtroom, he entered his plea. The judge fined him $10,000 and honored the plea agreement whereby Agnew received no jail term, but only three years unsupervised probation. For agreeing to cooperate with the prosecution, Matz and Childs were not prosecuted.

These events took place against the backdrop of one of the most intense government crises in U.S. history. Although Nixon and Agnew had been reelected in a landslide in the 1972 election, the Watergate scandal hung over the administration. Shortly after the events of this case, the Watergate scandal intensified, culminating in the resignation of Richard Nixon from the presidency.

PROFESSIONAL SUCCESS: LOOKING FOR A JOB

Many ethical issues arise in the course of looking for a job. Even though as you approach graduation you are still an "amateur," ethical and professional behavior is expected during your job search. There are many ways to be unethical in searching for a job, such as exaggerating or falsifying your resume, or overstating expenses when getting reimbursed for an interview trip.

Other, less obvious, ethical concerns can occur during interview trips. For example, suppose you have had an on-campus interview with a large corporation. After the interview you have decided that you aren't really interested in this company. The company calls you later and asks you to come to company headquarters in Cleveland for a plant visit. You have a friend in Cleveland who you would like to visit. Is it acceptable to go on the plant trip? Why? Does the situation change if the plant trip is to Hawai'i? Does it change if your interest level in the company is low, but you honestly feel that you could be persuaded?

How do you decide what is acceptable during your job search? The easiest thing to do is to honestly discuss your plans with the recruiter. If she feels that what you want to do isn't acceptable, then you shouldn't do it. If, however, your plans are acceptable to the company then you can proceed. In addition, the ethical analysis and problem solving methods that we developed in chapter 4 and have applied to cases throughout this book, are equally applicable to job searches.

PROFESSIONAL SUCCESS: CHEATING ON ASSIGNMENTS

The intense pressure to get good grades in college often leads to temptations to cheat on exams or assignments. Cheating is an issue that is likely to have arisen in educational settings even before you began your study of engineering. Of course the stakes become higher in a college or university setting, so the temptation to cheat might seem larger now than in high school. Cheating can take many forms, including copying someone else's work or using "cheat sheets" during an exam.

Although it can be analyzed using utilitarianism or rights and duty ethics, it is perhaps easiest to examine cheating using virtue ethics. As discussed in Chapter 3, honesty is a virtue. Honesty facilitates trust between individuals whereas dishonesty causes friction. People rarely want to associate with others who they feel don't behave fairly and can't be trusted. Cheating or falsifying work is a form of dishonesty. We should seek to enhance virtues such as honesty within ourselves and others, so virtue ethics clearly tells us that cheating is unethical.

As with many of the problems discussed in this book, you may be unsure whether something you are doing really is cheating. When faced with these types of problems, the same analysis and problem solving methods that are discussed in this book can be used to determine the correct course of action.

KEY TERMS

Line drawing
Flow charting
Bribery

REFERENCES

CHARLES E. HARRIS, JR., MICHAEL S. PRITCHARD, AND MICHAEL J. RABINS, *Engineering Ethics, Concepts and Cases*, Wadsworth Publishing Company, Belmont, CA, 1995.

MIKE W. MARTIN AND ROLAND SCHINZINGER, *Ethics in Engineering*, 2d. ed., McGraw-Hill, New York, 1989.

Magnetic Fields

TEKLA S. PERRY, "Today's view of magnetic fields," *IEEE Spectrum*, December 1994, p. 14.

Spiro Agnew

RICHARD M. COHEN AND JULES WITCOVER, *A Heartbeat Away: The Investigation and Resignation of Vice President Spiro T. Agnew*, Viking, New York, 1974.

New York Times, October 11, 1973. Numerous articles, starting with the front-page article about Agnew's resignation and his appearance in court. Articles leading up to this event can also be found in copies of the *New York Times* up to several weeks before this date.

Problems

1. Analyze the factual, conceptual, and moral issues in the following cases: a) the DIA case, b) the Pentium chip case, and c) the space shuttle *Challenger* case. What conclusion do you come to about the ethically correct choice that should have been made in each case?
2. Use the line-drawing technique described previously to analyze the actions of the following corporations: a) the 3Bs construction company in the DIA case and b) Paradyne in its dealings with the Social Security Administration. What conclusions do you draw about the actions of these companies?
3. Use flow charting to decide whether the space shuttle *Challenger* should have been launched. You can start at the beginning of the shuttle program and include some of the problems encountered in the development of the booster rockets, include consequences for NASA and Thiokol of delays in the launch, etc.
4. Use flow charting to analyze the actions of the Aberdeen Three. What conclusions can be drawn from this chart regarding their actions?
5. Use line drawing to assess whether the scenarios of bribery/gift giving under Examples in Section 4.6 are acceptable. What other examples can you think of to add to these scenarios?
6. Use flow charting to analyze whether the examples given in Section 4.6 are legitimate gifts or bribes. Be sure to indicate what consequences will flow from each decision.

Magnetic Fields

7. What does utilitarianism tell us about this case? What do rights and duty ethics tell us? Consider these questions from the point of view of a design engineer who must work on a product that might emit hazardous radiation. Which ethical theory applies best in this case? What does the code of ethics of the IEEE tell us about this case?
8. Analyze this case by determining the factual issues, determining the conceptual issues, and deciding which moral issues apply. Hint: This case is a perfect instance of what we discussed previously in this chapter when we said that the factual issues can be controversial.
9. If there are potential, but not well-understood, hazards in building a product, what are the future consequences of doing nothing—i.e., of making no changes in the design? Will warnings to the consumer suffice to get the designer off the hook? Must the product be engineered to be totally safe at all costs?
10. How can one best balance safety with economics in this case?
11. In their book *Ethics in Engineering*, Martin and Schinzinger state that "[e]ngineering, more than any other profession, involves social experimentation." How applicable is this statement to this case? Do you think that this statement is true in general?
12. In light of the results of various panels that indicate that there is no hazard associated with low-frequency magnetic fields, what should an engineer do today when designing products that will emit this type of radiation?

Spiro Agnew

13. Does the fact that paying government officials for receiving contracts seemed to be a commonplace business practice in Maryland at the time make this practice ethically acceptable?
14. What should an engineer do in the face of competition from others who are willing to resort to bribery?
15. What issues does this case raise regarding competitive bidding for engineering services? Would competitive bidding for the engineering contracts in Baltimore County have solved this problem? Keep in mind the issues brought up with regard to competitive bidding in the Paradyne case in Chapter 2.
16. What is the ethical status of a campaign contribution given to a politician to secure future business? Is this a bribe? Is it the same as a kickback? Perhaps line drawing would help answer this question.

5

Risk, Safety, and Accidents

On a sunny afternoon in May of 1996, Valujet Flight 592 took off from Miami International Airport, heading for Atlanta. Within minutes of leaving the runway, the DC-9's electrical systems started to fail and the cockpit and passenger cabin began filling with smoke. The pilots immediately called the Miami tower for permission to return and began to descend and turn back towards the airport. However, the situation worsened as fire started melting control cables and the pilots became overcome with smoke. The plane suddenly banked sharply and descended rapidly. The descent was so fast that the air-traffic control radar in Miami was no longer able to register an altitude for the airplane. Miraculously, the plane came out of its steep dive and leveled off, either through the efforts of the pilots or because the autopilot came back on. The airplane was now at only 1,000 feet above the ground. The air-traffic controllers in Miami radioed the pilots and attempted to send the aircraft to the closer airport at Opa Locka, Florida. Instead, Flight 592 rolled sharply to the right and, facing nose down, crashed into the Everglades. The two pilots, three flight attendants, and 105 passengers on board were killed instantly.

The subsequent investigation into this accident indicated that the fire was caused by the accidental firing of at

Sections

Objectives

After reading this chapter, you will be able to:

- Know the definitions of risk and safety.
- Discover different factors that affect the perception of risk.
- Study the nature of accidents.
- Know how to ensure that your designs will be as safe as possible.

least one of many chemical oxygen generators that had been removed from another Valujet airplane and were being carried back to Valujet headquarters in Atlanta. The heat generated by this canister caused a fire in the cargo hold beneath the cockpit that ultimately brought Flight 592 down. The investigation showed that these canisters were improperly secured and shouldn't have been on the airplane at all.

One of the most important duties of an engineer is to ensure the safety of the people who will be affected by the products that he designs. As we have seen, all of the codes of ethics of the professional engineering societies stress the importance of safety in the engineer's duties. As we will see later in this chapter, the cause of the Valujet accident wasn't a flaw in the airplane's design, but rather was attributed to a series of mistakes in the securing and handling of the oxygen canisters. What responsibility does the engineer have for ensuring that these types of mistakes are not made? How can products be designed to minimize the risk to the user? We will explore these questions in this chapter.

5.1 INTRODUCTION

No duty of the engineer is more important than her duty to protect the safety and well-being of the public. Indeed, the codes of ethics of the professional engineering societies make it clear that safety is of paramount importance to the engineer. In this chapter, we will look into safety and risk. We will also examine the nature of accidents and try to determine what the engineer's role is in preventing accidents and ensuring the safety of the public.

5.2 SAFETY AND RISK

At the core of many of the cases that we will study in this text are issues of safety and risk. As we have seen in the engineering codes of ethics, engineers have a responsibility to society to produce products that are safe. In discussing the Pentium chip case, we saw that there is an implied warranty of all products that they will perform as advertised—a bridge should allow automobiles to cross from one side of a river to the other, and a computer should correctly perform calculations. Similarly, there is an implied warranty that products are safe to use. Clearly, nothing can be 100% safe, but engineers are required to make products as safe as reasonably possible. Thus, safety should be an integral part of any engineering design.

Definitions

Safety is at the same time a very precise and a very vague term. It is vague because, to some extent, safety is a value judgment, but precise because in many cases, we can readily distinguish a safe design from an unsafe one. It is impossible to discuss safety without also including a discussion of risk. Risk is a key element in any engineering design; it is impossible to design anything to be completely risk free. How much risk is appropriate? How safe is safe enough? To answer these questions, we must first study the nature of safety and risk.

The American Heritage Dictionary defines risk as the possibility of suffering harm or loss. Risk is sometimes used synonymously with danger. The same dictionary defines safety as freedom from damage, injury, or risk. There is some circularity to these definitions: We engage in risky behavior when we do something that is unsafe, and something is unsafe if it involves substantial risk.

Although these definitions are precise, safety and risk are essentially subjective and depend on many factors:

1. Voluntary vs. involuntary risk. Many consider something safer if they knowingly take on the risk, but would find it unsafe if forced to do so. If the property values are low enough, some people will be tempted to buy a house near a plant that emits low levels of a toxic waste into the air. They are willing to assume the risk for the benefit of cheap housing. However, if a person already living near a plant finds that toxic fumes are emitted by the plant and he wasn't informed, the risk will appear to be larger, since it was not voluntarily assumed. This principle is true even if the level of emission is identical to that in the example of a person choosing to move near the plant.
2. Short-term vs. long-term consequences. Something that might cause a short-lived illness or disability seems safer than something that will result in permanent disability. An activity for which there is a risk of getting a fractured leg will appear much less risky than an activity with a risk of a spinal fracture, since a broken leg will be painful and disabling for a few months, but generally full recovery is the norm. Spinal fractures, however, can lead to permanent disability.
3. Expected probability. Many might find a 1 in 1,000,000 chance of a severe injury to be an acceptable risk, whereas a 50:50 chance of a fairly minor injury might be unacceptable. Swimming at a beach where there is known to be a large concentration of jellyfish would be unacceptable to many, since there would be a high probability of a painful, though rarely fatal, sting. Yet, at the same beach, the risk of a shark attack is low enough that it doesn't deter anyone from swimming, even though such an attack would very likely lead to death or dismemberment. It is important to remember here that the expected probability is only an educated guess.
4. Reversible effects. Something will seem less risky if the bad effects are ultimately reversible. This concept is similar to the short-term vs. long-term risk question discussed previously.
5. Threshold levels for risk. Something that is risky only at fairly high exposures will seem safer than something with a uniform exposure to risk. For example, the probability of being in an automobile accident is the same regardless of how often you drive. (Of course, you can reduce the likelihood of being in an accident by driving less often.) In contrast, studies have shown that low levels of nuclear radiation actually have beneficial effects on human health, while only at higher levels of exposure are there severe health problems or death. If there is a threshold for the effects, generally there will be a greater tolerance for risk.
6. Delayed vs. immediate risk. An activity whose harm is delayed for many years will seem much less risky than something with an immediate effect. For example, for several years now, Americans have been warned about the adverse long-term health effects of a high-fat diet. This type of diet can lead to chronic heart problems or stroke later in life. Yet, many ignore these warnings and are unconcerned about a risk that is so far in the future. These same people might find an activity such as skydiving unacceptably risky, since an accident will cause immediate injury or death.

Thus, whether something is unsafe or risky often depends on who is asked. Something that one person feels is safe may seem very unsafe to someone else. This creates some confusion for the engineer who has to decide whether a project is safe enough to be pursued. In making a decision, some of the analysis methods discussed in Chapter 4, especially line drawing and flow charting, can be used. Ultimately, it is up to the engineer and company management to use their professional judgment to determine whether a project can be safely implemented.

Safety and the Engineer

Since safety is an essential aspect of our duties as engineers, how can we be sure that our designs are safe? There are four criteria that must be met to help ensure a safe design. First, the minimum requirement is that a design must comply with the applicable laws. This requirement should be easy to meet, since legal standards for product safety are generally well known, are published, and are easily accessible.

Second, an acceptable design must meet the standard of "accepted engineering practice." You can't create a design that is less safe than what everyone else in the profession understands to be acceptable. For example, federal safety laws might not require that the power supply in a home computer be made inaccessible to the consumer who opens up her computer. However, if most manufacturers have designed their supplies so that no potentially lethal voltages are accessible, then that standard should be followed by all designers, even if doing so increases the cost of the product. A real-life example of this will be shown later when we consider the DC-10 case, in which an airframe was adapted from another design, but was not in accordance with the practice of other aircraft manufacturers at the time. This requirement is harder to comply with than the legal standard, since "accepted engineering practice" is a somewhat vague term. To address this issue, an engineer must continually upgrade her skills by attending conferences and short courses, discussing issues with other engineers, and constantly surveying the literature and trade magazines for information on the current state of the art in the field.

Third, alternative designs that are potentially safer must be explored. This requirement is also difficult to meet, since it requires a fair amount of creativity in seeking alternative solutions. This creativity can involve discussing design strategies with others in your field and brainstorming new alternatives with them. The best way to know if your design is the safest available is to compare it to other potential designs.

Finally, the engineer must attempt to foresee potential misuses of the product by the consumer and must design to avoid these problems. Again, this requires a fair amount of creativity and research. It is always tempting to think that if someone is stupid enough to misuse your product, then it's their own fault and the misuse and its consequences shouldn't bother you too much. However, an engineer should execute designs in such a way as to protect even someone who misuses the product. Of course, juries aren't always concerned with the stupidity of the user and might return a substantial judgment against you if they feel that a product was not properly designed. Placing a warning label on a product is not sufficient and is not a substitute for doing the extra engineering work required to produce a safe design.

Designing for Safety

How should safety be incorporated into the engineering design process? Texts on engineering design often include some variation on a basic multistep procedure for effectively executing engineering designs. One version of this process is found in [Wilcox, 1990] and is summarized as follows:

1. Define the problem. This step includes determining the needs and requirements and often involves determining the constraints.
2. Generate several solutions. Multiple alternative designs are created.
3. Analyze each solution to determine the pros and cons of each. This step involves determining the consequences of each design solution and determining whether it solves the problem.
4. Test the solutions.
5. Select the best solution.
6. Implement the chosen solution.

In step 1, it is appropriate to include issues of safety in the product definition and specification. During steps 2 through 5, engineers typically consider issues of how well the solution meets the specifications, how easy it will be to build, and how costly it will be. Safety and risk should also be criteria considered during each of these steps. Safety is especially important in step 5, where the engineer attempts to assess all of the tradeoffs required to obtain a successful final design. In assessing these tradeoffs, it is important to remember that safety considerations should be paramount and should have relatively higher weight than other issues.

Minimizing risk is often easier said than done. There are many things that make this a difficult task for the engineer. For example, the design engineer often must deal in uncertainties. Many of the risks can only be expressed as probabilities and often are no more than educated guesses. Sometimes, there are synergistic effects between probabilities, especially in a new and innovative design for which the interaction of risks will be unknown. Risk is also increased by the rapid pace at which engineering designs must be carried out. The prudent approach to minimizing risk in a design is a "go slow" approach, in which care is taken to ensure that all possibilities have been adequately explored and that testing has been sufficiently thorough. However, this approach isn't always possible in the real world.

Are minimizing risks and designing for safety always the more expensive alternatives? Spending a long time engineering a safer product may seem like a very expensive alternative, especially early in the design cycle before the product has been built or is on the market. This, however, is a very short-term view. As we will see in the Pinto case, the management at Ford took this short-term viewpoint and decided not to spend the extra money required to improve the design for the placement of the gas tank. A more long-term view looks at the possible consequences of not minimizing the risk. There is a great deal of guesswork involved here, but it is clear that any unsafe product on the market ultimately leads to lawsuits that are expensive to defend even if you don't lose and are very costly if you do lose. The prudent and ethical thing to do is to spend as much time and expense as possible up front to engineer the design correctly so as to minimize future risk of injury and subsequent criminal or civil actions against you.

Risk–Benefit Analysis

One method that engineers sometimes use to help analyze risk and to determine whether a project should proceed is called risk–benefit analysis. This technique is similar to the cost–benefit analysis that we discussed in Chapter 2. In risk–benefit analysis, the risks and benefits of a project are assigned dollar amounts, and the most favorable ratio between risks and benefits is sought. Recall that cost–benefit analysis is tricky because it is frequently difficult to assign realistic dollar amounts to alternatives. This task is especially difficult in risk–benefit analysis because risks are much harder to quantify and more difficult to put a realistic price tag on. Still, this can be a useful technique if used as part of a broader analysis and only if used objectively.

In doing a risk–benefit analysis, one must consider who takes the risks and who reaps the benefits. It is important to be sure that those who are taking the risks are also those who are benefiting. This consideration is fundamental to issues of economic justice in our society and can be illustrated by the concept of "environmental racism," which is the placing of hazardous-waste sites, factories with unpleasant or noxious emissions, etc. near the least economically advantaged neighborhoods. This practice is sometimes thought of as racism because in the United States, these types of neighborhoods are generally disproportionately occupied by minority groups. The only ethical way to implement risk–benefit analysis is for the engineer to ensure to the greatest extent possible that the risks as well as the benefits of her design are shared equally in society.

5.3 ACCIDENTS

Now that we have discussed some basic ideas related to safety and risk, it will also be useful to look at ideas on the nature of accidents and see how these ideas bear on our discussion of safety and the engineer's duty to society. There have been numerous studies of accidents and their causes, with attempts to categorize different types of accidents. The goal of this type of work is to understand the nature of accidents and therefore find ways to try to prevent them. Since the engineer's most important job is to protect the safety of the public, the results of this type of research have an impact on the engineering professional.

There are many ways in which accidents can be categorized and studied. One method is to group accidents into three types: procedural, engineered, and systemic [Langewiesche, 1998]. Procedural accidents are perhaps the most common and are the result of someone making a bad choice or not following established procedures. For example, in the airline industry, procedural accidents are frequently labeled as "pilot error." These are accidents caused by the misreading of an important gauge, flying when the weather should have dictated otherwise, or failure to follow regulations and procedures. In the airline industry, this type of error is not restricted to the pilot; it can also be committed by air-traffic controllers and maintenance personnel. Engineers must also guard against procedural problems that can lead to accidents. These problems can include failure to adequately examine drawings before signing off on them, failure to follow design rules, or failure to design according to accepted engineering practice. Procedural accidents are fairly well understood and are amenable to solution through increased training, more supervision, new laws or regulations, or closer scrutiny by regulators.

Engineered accidents are caused by flaws in the design. These are failures of materials, devices that don't perform as expected, or devices that don't perform well under all circumstances encountered. For example, microcracks sometimes develop in turbine blades in aircraft engines. When these cracks become severe enough, the blade can fail and break apart. Sometimes, this has resulted in penetration of the cabin by metal fragments, causing injury to passengers. Engineered failures should be anticipated in the design stage and should be caught and corrected during testing. However, it isn't always possible to anticipate every condition that will be encountered, and sometimes testing doesn't occur over the entire range of possible operating conditions. These types of accidents can be understood and alleviated as more knowledge is gained through testing and actual experience in the field.

Systemic accidents are harder to understand and harder to control. They are characteristic of very complex technologies and the complex organizations that are required to operate them. A perfect example of this phenomenon is the airline industry. Modern aircraft are very complicated systems. Running them properly requires the work of many individuals, including baggage handlers, mechanics, flight attendants, pilots, government regulators and inspectors, and air-traffic controllers. At many stages in the operation of an airline, there are chances for mistakes to occur, some with serious consequences. Often, a single, minor mistake isn't significant, but a series of minor mistakes can add up to a disaster. We will see this type of situation later in this chapter when we study the Valujet crash, in which several individuals committed a series of small errors, none of which was significant alone. These small errors came together to cause a major accident.

The airline industry is not the only complex engineered system in our society that is susceptible to systemic accidents. Both modern military systems, especially nuclear weapons, for which complicated detection and communication systems are relied on for control, and nuclear power plants with complicated control and safety systems, have documented failures in the past that can be attributed to this type of systemic problem. In our previous study of the accident at Bhopal, we see another case of a systemic accident.

What are the implications of this type of accident for the design engineer? Because it is difficult to take systemic accidents into account during design, especially since there are so many small and seemingly insignificant factors that come into play, it may seem that the engineer bears no responsibility for this type of accident. However, it is important for the engineer to understand the complexity of the systems that he is working on and to attempt to be creative in determining how things can be designed to avert as many mistakes by people using the technology as possible. As designers, engineers are also partially responsible for generating owner's manuals and procedures for the use of the devices they design. Although an engineer has no way of ensuring that the procedures will be followed, it is important that he be thorough and careful in establishing these procedures. In examining the Valujet accident, we will try to see how engineers could have designed some things differently so that the accident might have been averted.

APPLICATION: CASES

The Crash of Valujet Flight 592

Valujet was one of the generation of new discount airlines that sprang up as the result of airline deregulation in the 1980s. Based in Atlanta, it offered cheap fares to Florida and other popular destinations. Its cost savings were achieved in part by hiring other companies to perform many of the routine operations that keep an airline flying. For example, many major airlines perform aircraft maintenance themselves, work that Valujet hired a company named SabreTech to do. One of the jobs that SabreTech had been hired to perform for Valujet was the routine task of replacing oxygen-generator canisters in some of its DC-9s. This work was being performed at SabreTech's facility at Miami International Airport.

The oxygen canisters in the DC-9 are located above the passenger seats and are used to provide oxygen to the passengers through masks should the cabin pressure somehow be lost. The canisters contain a core of sodium chlorate, which is activated by a small explosive charge. This small explosion is initiated when the passenger pulls the oxygen mask toward herself. A chemical reaction within the canister liberates oxygen, which the passenger breathes through the mask. During use, the surface temperature of the canister can be as high as 500°F, which is normally not a problem, since the canister is mounted so that it is well ventilated. To ensure that they will operate properly when needed, the oxygen-generator canisters must be replaced periodically.

The Valujet maintenance rules made it clear that when the canisters are removed, a safety cap must be installed on them to ensure that the explosive charge is not inadvertently set off. Unfortunately, SabreTech didn't have any of these safety caps on hand while they were performing this work. Instead, tape was applied where the caps should have gone, and the canisters were placed in five cardboard boxes and left on a shelf in the hangar. However, two of the SabreTech mechanics marked on the paperwork that the caps had been installed and signed off on the job.

The five boxes of canisters sat on the shelf for several weeks, until a manager instructed a shipping clerk to clean up the area and get the boxes out of the hangar. Since the canisters were Valujet property, the shipping clerk prepared the boxes to be shipped back to Valujet headquarters in Atlanta. He rearranged the canisters, placing some of them end to end in the box, added some bubble pack on top, and sealed up the boxes. To this load, he also added tires, some of them mounted on wheels and probably filled with air. A shipping ticket was prepared describing the load as empty oxygen canisters (even though most of them were full) and tires. The load was delivered to Flight 592.

The Valujet ramp agent accepted the load despite the fact that Valujet was not certified to carry hazardous wastes such as empty oxygen generators, which contain a toxic residue from the chemical reaction. The flight's copilot, Richard Hazen, also looked at the load and the shipping ticket, but apparently didn't think that there was a problem with carrying this cargo. Together, the ramp agent and the copilot decided to put the load in the forward hold, which is underneath and behind the cockpit. The Valujet ground crew placed the tires flat on the bottom of the compartment and stacked the five boxes on top of the tires.

What happened to the plane after the cargo hold was loaded was reconstructed from the flight data recorder and the voice recorder, the "black boxes" that all planes are required to carry. Takeoff of Flight 592 was normal. But six minutes into the flight, there was a beep on the public-address system. At the same time, there was a sound like a chirp on the voice recorder. The flight data recorder indicated a pulse of pressure occurring simultaneously with these sounds. Accident investigators think that either during taxi or takeoff, one of the canisters was jostled and the explosive charge ignited. As the chemical reaction proceeded, the canister got extremely hot, especially since the canisters were in a box and were not ventilated as they are when mounted in the airplane. The chirping sound and the accompanying pressure surge were probably caused by one of the tires in the hold bursting due to the heat. At this point, the cardboard boxes and the tires were probably on fire. Suddenly, the plane's instruments started to indicate an electrical failure, presumably caused by the melting of some of the wiring that ran underneath the cabin floor.

As smoke filled the cabin, the pilot, Candalyn Kubek, struggled to regain control of her aircraft. Desperate radio messages were sent to air-traffic control in Miami, where controllers tried to route the plane back to Miami and, finally, to a closer airport. The pilots were unable to control the plane. It banked sharply to the right and dove nose first into the Everglades. All 110 persons aboard were killed.

This case seems to be a perfect example of a systemic accident. There were many small mistakes made by several people:

- The proper safety caps should have been installed.
- Although the safety caps were not installed on the oxygen canisters, had they been packed properly, this situation might not have been a problem.
- The ramp agent, who was trained to identify improper and hazardous cargo, should not have let these boxes on the airplane.
- The copilot, similarly trained, should also have refused to carry this cargo.
- Something that generates such intense heat should not have been put in such close proximity to a tire, which burns with very acrid and thick smoke.
- The cargo compartment should have had heat and smoke detectors to give the pilots advanced warning of trouble in the hold.

By themselves, none of these lapses should have led to the crash. However, the convergence of all these mistakes made the accident inevitable.

In the aftermath of this accident, the two SabreTech mechanics were fired for falsifying the paperwork. Valujet's entire fleet was grounded for several months. Valujet eventually returned to business later in 1996, but ultimately changed its name to AirTran. As a result of this crash, the FAA began to require airlines to install heat and smoke detectors in the cargo holds of all airplanes.

The Collapse of the Hyatt Regency Kansas City Walkways

In the 1970s, it became popular to design upscale hotels with large atriums, some extending the entire height of the hotel. This feature helped create very dramatic architectural spaces in the hotel lobbies and is still often seen in hotel design today. Many of these designs also included walkways suspended over the atrium. One hotel using this design is the Hyatt Regency Kansas City. Development of this hotel began in 1976, and construction was completed in the summer of 1980. One year later, in July of 1981, during a dance party in the atrium lobby, some of the walkways on which people were dancing collapsed onto the crowded atrium floor, leaving 114 people dead and 185 people injured.

The development of the Hyatt Regency Kansas City was initiated in 1976 by Crown Center Redevelopment Corporation, which hired Gillum-Colaco, Inc. of Texas as the consulting structural engineers. Gillum-Colaco worked closely with Crown Center Redevelopment and the project architects to develop the plans and create the structural drawings and specifications. Construction on the hotel began in 1978. Gillum-Colaco didn't actually perform the structural engineering for this project, but rather subcontracted this work to its subsidiary, Jack. D. Gillum and Associates, Ltd.

The general contractor for the project was Eldridge Construction Company, which hired Havens Steel Company as the subcontractor for fabrication and erection of the atrium steel. The original design called for the walkways to be hung from rods connected to the atrium ceiling. There would be two walkways connected to each rod by separate nuts. (See Figure 5.1.) Implementation of this design required that the rods be threaded for most of their length, which would greatly increase the cost of the rods. Havens suggested a change in the design that would avoid the requirement for threading long pieces of rod. It is not uncommon for a subcontractor to suggest changes in a structure, especially if the changes can lead to cost savings or easier fabrication. The changed design, shown in Figure 5.2, required that only a shorter section near the ends of the rods be threaded. In the original design, each of the nuts only supported the weight of one floor of the walkway. Unfortunately, in the revised design, some of the nuts supported the weight of both walkways, effectively doubling the load on the nuts. Gillum and Associates later claimed never to have seen any documents related to this change. Nor, they claimed did anyone from Havens or Eldridge contact them about this change. However, drawings indicating these changes were stamped with Gillum's seal in February of 1979.

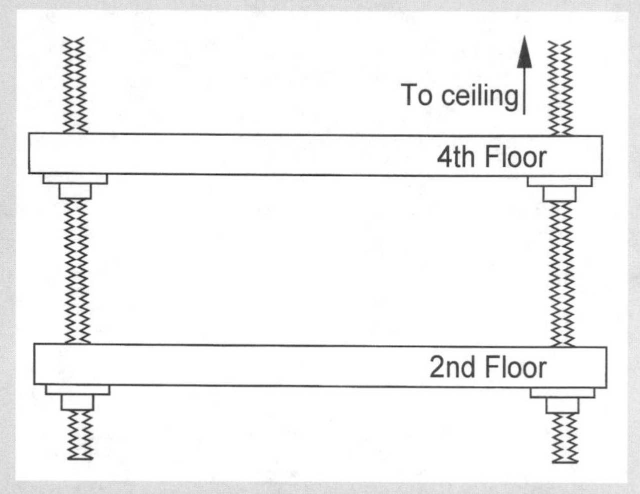

Figure 5.1. Schematic drawing of the second- and fourth-floor walkway supports as originally designed. The nuts beneath the fourth-floor walkway only support that walkway.

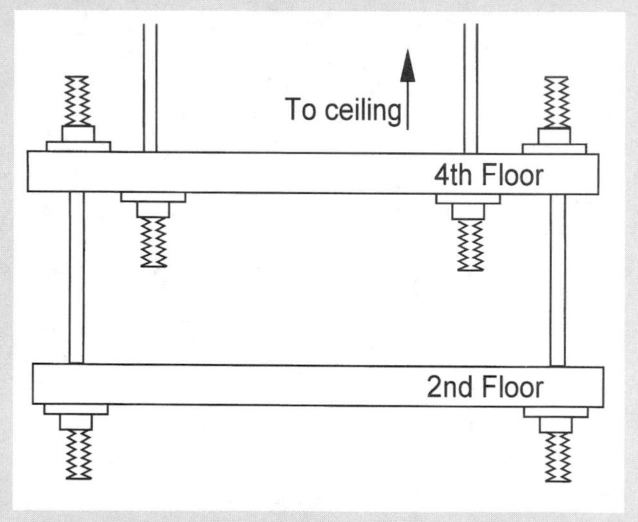

Figure 5.2. Schematic drawing of the second- and fourth-floor walkway supports as built. The nuts beneath the fourth-floor now support the weight of both the fourth- and second-floor walkways—twice the load of the original design.

In October of 1979, during construction, part of the roof of the atrium collapsed. There were investigations of this by Gillum and Associates and by an independent engineering firm. Reports were sent to the owners and architects assuring them that the atrium design was safe. In July of 1980, the hotel opened for business. In July of 1981, during a dance, many people were dancing on the second- and fourth-floor walkways. The load caused by the large number of people on the walkways and by the slight swaying that might have resulted from the dancing helped cause the failure of the connections that held up the walkway, resulting in the deaths and injuries.

In the wake of the accident, an investigation was conducted by the Missouri Board of Architects, Professional Engineers and Land Surveyors. This investigation resulted in charges of negligence, incompetence, and misconduct on the part of Gillum and Associates and its parent company. The report indicated that the original design was only marginally acceptable and didn't conform to the Kansas City building code. As originally designed, the walkways would only have had approximately 60% of the capacity required by the code. The changes initiated by Havens and approved by Gillum made this situation even worse.

Gillum and Associates was also found to be negligent in its investigation of the atrium collapse during construction and was found to have placed too much reliance on Havens. As a result of this accident, Jack Gillum lost his license to practice engineering, and Gillum and Associates lost its license as an engineering firm.

The Ford Pinto Exploding Gas Tank

In the late 1960s, Ford Motor Company set out to produce a subcompact automobile that would compete with the Japanese subcompacts that were becoming increasingly popular in the United States. The president of Ford, Lee Iacocca, set an ambitious two-year agenda for the design and initial production of the Pinto. This schedule meant that the car would come to market in 25 months rather than go through the average 43-month design cycle at Ford. Iacocca wanted the car to be introduced in the 1971 model year with a suggested retail price of no more than $2,000 and with a weight of less than 2,000 pounds. Cost and weight constraints are something that automotive engineers always face, and Ford engineers certainly knew how to deal with them. However, the fast design cycle presented great challenges to the engineering staff.

One of the myriad decisions that confronted the engineers was the placement of the gas tank. The tank could have been placed over the differential, where it would have been somewhat safe in the event of a rear-end collision. However, considerations of trunk space and manufacturing cost dictated that the tank be placed farther back, between the differential, which had several exposed bolt heads, and the rear bumper. In this position, a rear-end collision might push the gas tank forward into the differential, where the exposed bolts could rupture the tank, possibly leading to a fire or explosion.

Ford engineers knew that the Pinto's gas-tank design was susceptible to explosions from rear-end collisions, mainly from previous experience with the Capri, a European car produced by Ford on which this problem had occurred and had been fixed. In rear-end collision tests, a Capri with a rear-mounted tank was susceptible to gas-tank rupture in impacts as low as 20 miles per hour. This problem was corrected on the Capri by mounting the tank over the differential, which resulted in no tank rupture after rear-end impacts as high as 30 miles per hour. Ford actually noted the Capri's safer fuel tank location in its advertising.

An alternative to moving the Pinto's tank would have been to install a part costing Ford $6.65 that would have protected the tank from rupture due to the exposed bolts and would have gone a long way toward reducing the possibility of injury as a result of a rear-end collision. However, Ford did not choose to implement this alternative. In Ford's defense on this point, although $6.65 seems trivial, given the number of Pintos that it expected to produce, Ford estimated that over the production run of the Pinto, adding this part would have ultimately cost the company over $20 million [DeGeorge, 1981]. It is also important to note that at the time the Pinto was designed, there were no applicable federal safety standards regarding the placement of gas tanks in automobiles. There were also no regulations regarding the safety of gas tanks in rear-end collisions, although Ford engineers were aware that such standards were on the horizon and that the Pinto would not meet these new federal regulations.

It was mentioned at the beginning of Chapter 1 that an accident occurred on August 10, 1978, in which a 1973 Pinto was hit from behind near Goshen, Indiana. The gas tank caught fire, and Judy Ulrich, age 18, her sister Lynn Ulrich, age 16, and their cousin Donna Ulrich, 18, died of burns received in the fire. On September 13, 1978, Ford was indicted and charged with criminal negligence. Ultimately, Ford was not convicted, but this case and many other similar Pinto accidents cost Ford millions of dollars in legal settlements to accident victims. Ford also suffered untold damage to its reputation for producing well engineered, safe automobiles. During the litigation resulting from these accidents, Ford claimed that the Pinto met applicable federal standards and was as safe in this regard as other subcompact cars. However, similar models didn't seem to be susceptible to this problem as the Pinto was. In fact, Ford was eventually forced to recall the Pintos in order to install the $6.65 part. Including labor, the cost to do this procedure far exceeded the cost of the part. No other similar subcompacts made by other manufacturers were recalled for this reason.

There are some other interesting facts that bear on this case. At the height of the controversy over the Pinto, a Ford vice president bought one for use by his daughter. This event was widely reported and was seen as a vote of confidence in the safety of the product. In addition, when the news got out about the problems with the Pinto, many rental companies tried to sell off their fleets of Pintos. Interestingly, they had no trouble selling them to a presumably knowledgeable public. Apparently, given the right price, people were willing to take the risk involved in owning this car.

One would think that after this sort of well-publicized case, the automobile industry would be more careful about gas-tank placement. But as shown in the early 1990s, this is not the case. For example, GMC started selling its pickup truck with side gas tanks that were outside the structural frame of the vehicle. This design led to accidents in which the tanks ruptured and the ensuing fires caused severe injuries. One such accident led to the death of a teenager in Georgia. Although the case was sensationalized by NBC, which later admitted to faking the explosion of gas tanks for their cameras (yet another area ripe for an interesting ethical discussion), the fact remained that the design was flawed and didn't meet the standard of accepted engineering practice. GMC has subsequently redesigned this truck to be more safe.

The Failure of the Teton Dam

On June 5, 1976, the Teton dam in Idaho failed, releasing millions of gallons of water into the Snake River. The subsequent downstream flood caused 14 deaths and damage that was estimated at between $400 million and $1 billion. The Teton dam was designed and built by the Bureau of Reclamation, which had extensive experience building dams, including ones of this type. Reclamation had been responsible for building over 300 dams, including the Hoover and Grand Coulee dams. The Teton dam was a conventional earthfill dam, of which Reclamation had already built approximately 250. What was not conventional in this case was the quality of the rock in the surrounding canyon walls, which led to the failure of the dam before it had even been completely filled.

A typical earthfill dam, such as the Teton dam, consists of a core that is a mound of fine silt, compacted in order to make it impervious to the flow of water. The core accounts for about half of the volume of the dam. The silt core is covered with sand, gravel, and cobbles. Layers of other materials, such as earth and rocks, form additional shells; in all, there were five layers of different materials in the Teton dam. Completed, the dam was 305 feet high.

The condition of the surrounding canyon walls imposed some unique conditions on the design and construction of the Teton dam. The rock in the walls was highly fractured, providing passages through which water can pass. This situation is a problem for earth dams, since water that is being held back can infiltrate the rock and pass into the dam, causing erosion of the earth in the structure. Once erosion begins, further water can leak into the dam, increasing the erosion and increasing the flow of water in an increasing cycle, leading to catastrophic failure. The highly fractured rock found at the Teton site could easily have caused this type of water flow and dam failure. During the drilling of test holes in the rock, most of the cracks that were discovered were very small. However, when the foundation of the dam was excavated, numerous large cracks were discovered in the canyon walls. In fact, one of these fissures was four feet wide and allowed an inspector to walk through it about 100 feet both upstream and downstream.

Normally, cracks in canyon walls will not prevent the construction of an earthfill dam, since there are methods by which these cracks can be neutralized. At the Teton site, Reclamation cut trenches in the top of the canyon walls on both sides of the dam. The trenches were 70 feet deep and extended back 1,000 feet into the canyon walls. Digging these trenches removed much of the most severely fragmented rock. At the bottom of the trenches, a series of holes was drilled in three parallel lines. These holes extended as much as 300 feet deep to well below the base of the dam. Grout was pumped into these holes to seal any remaining cracks in the canyon walls. A single row of grout-filled holes was also installed on the floor of the canyon beneath where the dam embankment would be. After grouting, the trenches on the canyon walls were filled with silt, the same material that was in the

The aftermath of the failure of the Teton Dam in June, 1976. Many areas downstream were inundated with water when the dam collapsed. Photo by Glade Walker / U.S. Bureau of Reclamation.

core of the dam. In theory, all of this resulted in a barrier that was impermeable to water. Any water from the dam that escaped into the rock should have had a long and impossible route back to the face of the dam and would instead have been likely to emerge from the rock somewhere downstream of the dam. Unfortunately, all of these efforts did not work. As the water in the dam was reaching capacity for the first time, the dam failed.

In the wake of the disaster, many studies and investigations were initiated. One of the investigations was performed by a panel of nongovernmental experts—civil engineers and dam builders—appointed by the Secretary of the Interior (the Bureau of Reclamation's parent organization) and the governor of Idaho. Hampering these investigations was the fact that most of the dam, and certainly the part that failed, was swept downstream and destroyed and so was unavailable for analysis. The panel concluded that water got through the barriers implanted in the canyon wall either by passing through a portion of rock that was not plugged by the grout or by traveling directly through cracks in the silt channel caused by differential strains or hydraulic fracturing. Crossing the trench in this way, the water cut a hole through the silt, then flowed through cracks in the rocks to the core of the dam. There, it dug channels in the core materials, weakening the dam and leading to the collapse of the entire structure. The panel found that "the failure was caused not because some unforseeable fatal combination existed, but because the many combinations of unfavorable circumstances inherent in the situation were not visualized, and because adequate defenses were not included in the design" [Boffey, 1977].

Among the findings of the panel [Boffey, 1977]:

- There was too much reliance on the grout curtain. Some leakage is inevitable, so the design should have included provisions to reduce the consequences of a leak.

- The panel found that the silt used in the core and the trench fill was of inadequate quality, increasing the potential for erosion and cracking. In addition, this erodible silt was placed next to highly cracked rock in the canyon wall and floor, where water was certain to get at it.
- The trenches in the canyon walls were narrow and steep, resulting in stress patterns that encouraged cracking, hydraulic fracturing, and erosion of the silt used to fill the trenches.
- There was inadequate provision made for handling leakage. Normally, the gravel and rocks surrounding the dam's core are supposed to carry away any water leakage. That didn't seem to occur at the Teton dam.
- There was inadequate instrumentation to monitor conditions in the dam's embankment and the surrounding canyon walls. There was no clue that there was a problem until just hours before the collapse, when it was no longer possible to do anything. The panel felt that had there been better instrumentation, the early signs of the failure might have been detected and remedies might have been applied before the disaster.

The DC-10 Case

In July of 1970, the McDonnell-Douglas corporation, one of three major commercial-airliner manufacturing companies in the United States, was pressure-testing the new DC-10, a wide-body aircraft designed to compete with the soon-to-be introduced Boeing 747 and Lockheed L-1011. At the time, Boeing had an edge on the competition, since it was converting to civilian use an existing design for a large military cargo plane. It had not won the contract to build the cargo plane, but with little modification could have adapted the design to the commercial-airliner market. Turning an existing design into a new civilian airliner is clearly easier than starting from scratch, so Boeing had a large advantage in the potentially huge widebody-aircraft market. Lockheed was well along with its design as well. During the pressure testing of the DC-10, the cargo door on the prototype blew out and the floor of the passenger compartment buckled.

In 1972, after introduction of the DC-10 airliner, an American Airlines DC-10 flying over Windsor, Ontario suffered an accident similar to the one experienced during testing. The cargo door failed in flight, collapsing the cabin floor. In the DC-10, the electrical and hydraulic lines that are used to control the plane were routed under this floor. In this accident, several of these hydraulic lines were severed. Fortunately, the pilot was still able to control the plane and brought it to a safe landing in Detroit.

In March of 1974, the same type of accident occurred again, this time on a Turkish Airlines DC-10 carrying 346 passengers. At 10,000 feet over the suburbs of Paris, the cargo door failed and the floor of the passenger compartment collapsed. This time, all of the hydraulic and electrical connections were severed, rendering the airplane uncontrollable. The plane crashed, and everyone on board was killed.

Two aspects of the aircraft design were blamed for these accidents: the cargo-door latching system and the floor structure. Designers had considered both hydraulic and electric latches to secure the cargo door. A manual latching system was not feasible, since the cargo door was extremely large. The hydraulic and electric latches considered by the McDonnell-Douglas engineers had very different failure mechanisms. Briefly, if the hydraulic latch didn't close completely, it would fail when relatively small amounts of internal pressure built up on the door. Thus, the door would pop open soon after takeoff at relatively low altitudes, where the pressure difference between the cabin and the cargo hold would not be great enough to cause the floor to buckle, and the plane could still safely return to the airport [French, 1982].

On the other hand, if the electric latch hadn't closed completely, it would stay stuck until a much greater pressure had built up inside the aircraft. This situation would lead to a high-altitude blowout with catastrophic results. However, the DC-10 designers chose the electric latch for sound engineering reasons: It was lighter and had fewer moving parts than the hydraulic latch.

In designing the airframe, McDonnell-Douglas chose to make the DC-10 much like the older DC-8 and DC-9, two very successful and safe aircraft. The DC-10 engineers were constrained by management to utilize the existing airframe technology, which was not necessarily adequate for an aircraft the size of the DC-10.

Indeed, both Boeing and Lockheed had made several advancements in their airframe structural designs for the 747 and the L-1011, respectively [French, 1982]. These design advances were not proprietary, so the accepted engineering practice for building a jumbo jet was significantly different from what McDonnell-Douglas was doing [Eddy, Potter, and Page, 1976]. The structural integrity of an airliner is important, especially since the hydraulic control systems that operate the control surfaces of the aircraft must run somewhere through the airframe. Interestingly, both Lockheed and Boeing used four parallel redundant hydraulic systems, any one of which was capable of flying the aircraft. To save cost, McDonnell-Douglas chose to utilize only three parallel redundant systems. In addition, Boeing had chosen to route the control lines through the ceiling above the cabin, where they were not susceptible to damage when cabin pressure was lost.

The dismal safety record of the DC-10 continued in May of 1978, when an American Airlines DC-10 taking off from O'Hare International Airport in Chicago crashed when one of the underwing engines tore away from its support and fell to the ground. As it separated from the wing, the engine ripped through the hydraulic lines that ran through the wings, leaving the pilots with no means for adjusting the control surfaces on that side of the aircraft. The pilots were unable to bring the airplane under control, and the plane crashed, killing everyone on board.

Officially, the cause of the crash was improper maintenance procedures by airline personnel. When the engines were removed for maintenance, they were not replaced using the correct procedure, causing minuscule cracks to form in the engine supports, leading to the failure. However, some blame can also be attributed to the DC-10's design. The triply redundant hydraulic lines that activated the wing's control surfaces were all located in the leading edge of the wing. When the engine tore away from its mount, it damaged all three systems. In contrast, the Boeing and Lockheed design called for four redundant hydraulic lines that were more safely spaced throughout the wing [Newhouse, 1982].

Clearly, some fault for these accidents belongs with the design engineers, although in part, their actions are the result of corporate policies. The corporation had a very conservative design culture that neither expected innovation nor desired to be at the leading edge [French, 1982]. In addition, McDonnell-Douglas was not doing well financially, and so there were economic constraints and competitive time constraints put on the engineers in executing the design. Thus, there seems to have been some collective corporate responsibility for these accidents.

There was also some individual responsibility for the accidents. Following the near-disaster in 1972, McDonnell-Douglas agreed with the FAA that the cargo-door latching system should be modified and began work to refit the latches on the existing fleet. The Turkish Airlines plane was sent to McDonnell-Douglas' plant in California in July of 1972. The plant records show that three inspectors stamped the maintenance records for this aircraft to signify that the door modifications had been completed. However, the work had not actually been done. The inspectors had either very weak excuses or no explanation at all about how their stamps appeared on the records even though the work had not been performed. Clearly, there is no excuse for not paying close attention to inspection details, especially on a critical system where human life is at stake. The fact that three inspectors failed to notice that the modifications had not been made as required indicates that McDonnell-Douglas had lax oversight of the inspection and modification process and is also responsible for the work not being performed.

In the aftermath of the 1978 Chicago crash, all DC-10s were grounded for rework and inspection. After they were recertified as airworthy by the FAA, the flying public was understandably loathe to fly on the DC-10. In fact, partly in a marketing effort to overcome these problems, McDonnell-Douglas eventually renamed its entire line of aircrafts with a new "MD" designation. Essentially the same plane, though redesigned for better safety, the "new" DC-10 became the MD-11. In 1997, Boeing merged with McDonnell-Douglas and has absorbed its former competitor's commercial aviation business into its own organization.

Cellular Phones and Automotive Safety

In October of 1993, a Ford Explorer was traveling on a county highway in Suffolk County, New York. The Explorer was equipped with a cellular phone that was mounted on the transmission hump between the front seats. While using the phone, the driver took her eyes off the road. The vehicle crossed over the center dividing line and struck an oncoming car head on. Three members of a family riding in the other car were severely injured and required extensive hospitalization. As a result of this accident, the victims sued the manufacturer of the cell phone, the company that made the mounting bracket for the phone, and the shop that installed the bracket and the phone.

This case is just one of many traffic accidents that have been caused by drivers whose attention was diverted by using cellular phones while driving. It seems intuitively obvious that using a cell phone while driving is dangerous, but how big is the risk? Surprisingly, there is little data on this since most police departments do not require accident investigators to gather information on whether a cell phone was in use during the time leading up to an accident. However, in a study published in early 1997, two Canadian researchers answered this important question. This study was motivated by an incident that happened to one of the researchers, a physician at a medical school in Toronto. He had returned a call to a patient. The number he dialed turned out to be a cell phone, which the patient answered while driving. During their brief conversation, the patient was in an accident, leading the physician to initiate a study on just how dangerous cell-phone use during driving can be.

The study was performed by looking at records of several hundred accidents in which cell phones were present in the car. With the permission of the drivers, the researchers obtained the cellular-phone records of the drivers to see whether they were using the phones at the time of the accidents. Their results indicate that the risk of being involved in an accident is four times greater when the driver is using a cell phone. For comparison, the researchers pointed out that a driver whose blood alcohol content is 0.10% has the same increased risk of being in an accident. This blood alcohol level is above the legal limit in most states in the United States. So, using a cell phone in a car seems to be just as risky as driving drunk!

In response to these types of problems, cellular-phone use while driving has been banned in Brazil, Israel, and Sweden. There has been talk of similar bans in several states in the United States, but as of early 1998, no state has enacted such a ban. In the absence of any regulations, it is estimated that by the year 2000, between 0.6 and 1.2% of all traffic accidents will be attributable to cell-phone use, causing between two and four billion dollars in damage yearly. The number of accidents and ensuing damage is held somewhat in check by the relative expense of using cell-phones—most users limit their time on the phone. If prices were to come down, the incentive to limit use would be gone, and accident rates would be expected to rise sharply above these predictions.

What responsibility do engineers have regarding this problem? Cell phones and the means for mounting them in automobiles are designed by engineers, so certainly some responsibility for ensuring that they can be used safely resides with the design engineer. An obvious way to make these phones safer to use is to design them so that they can be operated hands free. Indeed, this approach has been tried by several manufacturers, relying chiefly on speech recognition software and on digital voice synthesis. However, this doesn't necessarily solve the problem. The same Canadian study discussed previously looked into whether hands-free cell phones were safer to operate. Although the data were much more limited, the indication was that the accident risk is identical for hands-free phones compared to the more conventional types requiring significant driver attention to dialing. Why is this?

The researchers speculate that the key issue in accidents involving cell phones is the diversion of the driver's attention while talking on the phone and not necessarily the loss of the use of one hand during dialing. This result, if true, does not match our intuitive sense of the hazards of cell phones. After all, often there is another person in the car talking to the driver. Talking to a passenger would seem to be as diverting as talking on the phone. The difference appears to be that when there is a passenger and the conditions require added driver attention, the passenger intuitively knows to stop talking or changes to less stressful topics. Someone talking to the driver on a cell phone can't know what the traffic is like and will thus not know when to stop talking.

It was reported in 1997 that Microsoft Corporation was developing a dashboard-mounted computer that would allow drivers to check their e-mail and surf the Internet. Naturally, it plans to use voice recognition and speech synthesis for communication between the driver and the computer. This system poses yet another threat to driver attention and highway safety.

In defense of cell phones used in autos, it should also be pointed out that there are some ways in which safety can be enhanced by a cell phone. In the Canadian study, nearly half of the people who had been in accidents used the cell phones to call for help.

KEY TERMS

Safety | Risk | Procedural, engineered, and systemic accidents

REFERENCES

WILLIAM LANGEWIESCHE, *The Lessons of Valujet 592, The Atlantic Monthly,* March 1998, pp. 81–98.

ALAN D. WILCOX, *Engineering Design for Electrical Engineers,* Prentice Hall, Englewood Cliffs, NJ, 1990.

Valujet 592

WILLIAM LANGEWIESCHE, *The Lessons of Valujet 592, The Atlantic Monthly,* March 1998, pp. 81–98.

Kansas City Hyatt Regency Walkway Collapse

EDWARD O. PFRANG AND RICHARD MARSHALL, "Collapse of the Kansas City Hyatt Regency Walkways," *Civil Engineering-ASCE,* July 1982, pp. 65–68.

Ford Pinto

RICHARD T. DEGEORGE, "Ethical Responsibilities of Engineers In Large Organizations: The Pinto Case," *Business and Professional Ethics Journal,* vol. 1, no. 1, Fall 1981, p. 1.

Chicago Tribune, October 13, 1979, p. 1.

Teton Dam

PHILLIP M. BOFFEY, "Teton Dam Verdict: A Foul-up by the Engineers," *Science,* vol. 195, 21 January 1977, pp. 270–272.

DC-10

PAUL EDDY, ELAINE POTTER, AND BRUCE PAGE, *Destination Disaster,* New York, New York Times Book Co., 1976.

PETER FRENCH, "What is Hamlet to McDonnell-Douglas or McDonnell-Douglas to Hamlet: DC-10," *Business and Professional Ethics Journal,* vol. 1, no. 2, 1982, pp. 1–19.

PETER NEWHOUSE, "The Aircraft Industry," *The New Yorker,* June 14, 21, 28 and July 5, 1982.

Cell Phones

New York Times, February 13, 1997, Section A, p. 30.

New York Times, April 25, 1997, Section B, p. 23.

New York Times, June 30, 1997, Section D, p. 9.

Problems

1. Think of some type of risky or unsafe behavior in which you have participated. What made it seem unsafe? Why did you do it anyway? What does this tell you about your role as an engineer?
2. In what ways can the accident at Bhopal be described as procedural, engineering, or systemic?
3. How would you classify the space shuttle *Challenger* accident? Why?

Valujet 592

4. Is this accident an engineered, procedural, or systemic accident? Is it partly all three?
5. What does this accident tell us about how systemic accidents can be avoided?
6. How might the oxygen canisters have been engineered to prevent such an accidental firing? Is there a better way than safety caps to secure these canisters?
7. Should smoke and heat detectors have been installed in the cargo holds?

Kansas City Hyatt Regency Walkway Collapse

8. What type of accident was this? How could the accident have been avoided?
9. Ultimately, who was responsible for checking the drawings and approving changes?
10. What responsibility did Havens have for the collapse?
11. Who is responsible for ensuring that the applicable building codes are followed?
12. What responsibility does an engineer have for checking and ensuring that what is in the drawings is what actually goes into the building?

Ford Pinto

13. Was the design for gas-tank placement in the Pinto acceptable engineering? Does the experience with the Capri have an impact on this answer?
14. Did the Pinto design meet acceptable engineering standards? Is the fact that the design did not violate federal standards important?

15. Is this case an example of an engineering failure or a management failure?
16. It seems that much of the blame in this case lies with management rather than with the engineers. What are the responsibilities of the engineers who are aware of a problem before a final design is made and manufacturing begins?
17. What are the responsibilities of the engineers after the decision has been made?
18. Should the engineers have refused to design the Pinto on such a tight schedule?
19. What managerial changes or other corporate considerations could head off this type of problem?
20. Can it be said that by not making the change in the design of the Pinto, Ford saved the $6.65 per car and the consumers took the increased risk? What are the ethical implications of this?
21. Use line drawing to analyze this case from the perspective of Ford engineers or management. Did Ford do the right thing?
22. Make a flow chart that might have helped Ford engineers and managers analyze their decisions regarding gas-tank placement in the Pinto. What would have been the best choice?

Teton Dam

23. Many critics of the Bureau of Reclamation contend that the dam should never have been built, given the highly cracked nature of the rock in the surrounding canyon walls. Should the dam have been built under these circumstances?
24. Some critics charged that once the momentum had been built on this project, there was no stopping it, regardless of the problems that became evident during construction. Can all problems be overcome with more engineering? Should all problems be overcome with more engineering? Should new methods have been employed to solve the fissured rock problem, rather than relying on technology (the grout curtains) that Reclamation had experience with at other less fissured sites?
25. Seepage and leakage of water around this type of structure is probably inevitable. What precautions should engineers take to ensure that an inevitable problem is managed?
26. How much instrumentation and monitoring is enough in this type of situation?
27. Should this accident be classified as procedural, engineered, or systemic?
28. Draw a flow chart of the decisions made by the engineers as they designed the Teton dam. Where do you think they went wrong?

DC-10

29. Should engineers have refused to build the DC-10 on such an accelerated schedule?
30. Should the design engineers have insisted on using state-of-the-art designs even if doing so went against the corporate culture?
31. In the aftermath of the door blowout on the prototype, what should McDonnell-Douglas have done?
32. Given that McDonnell-Douglas management knew that there was a problem with the cargo door as early as 1970, when the prototype failed, and knew that there could be catastrophic failure of the airplane's superstructure, leading to an accident, what ethical responsibility for the accidents does it have?
33. The hydraulic latching system seems to be more "fail safe" than the electric system. What are the ethical implications of using the electric system in the DC-10?
34. What were the responsibilities of the inspectors who oversaw the modification of the doors? What responsibility does McDonnell-Douglas have to ensure that inspections are performed properly?
35. Analyze this case from the perspective of risk–benefit analysis. What conclusion would a reasonable manufacturer take? In assessing the risk, be sure to keep in mind that there is some financial risk to the company, but also some personal risk to the people flying on the airplane. Is it reasonable to expect passengers to understand the risks involved? Even though flying is acknowledged to be a risky endeavor?

Cell Phones

36. What responsibility does the design engineer have to ensure that the products he designs are safe to use? A cell phone can be safely used by stopping the car while talking, but not everyone does this. Is this the engineer's fault?
37. In what ways could cell phones be made safe to use in an automobile?
38. On balance, is the use of cell phones in automobiles a safety risk or a safety enhancement?

The Rights and Responsibilities of Engineers

In the early 1970s, work was nearing completion on the Bay Area Rapid Transit (BART) system in the San Francisco Bay metropolitan area. The design for BART was very innovative, utilizing a highly automated train system with no direct human control of the trains. In the spring of 1972, three engineers working for BART were fired for insubordination. During the course of their work on the project, the three had become concerned about the safety of the automated control system and were not satisfied with the test procedures being used by Westinghouse, the contractor for the BART train controls.

Unable to get a satisfactory response from their immediate supervisors, the engineers resorted to an anonymous memo to upper management detailing their concerns and even met with a BART board member to discuss the situation. The information on the problems at BART was leaked to the press by the board member, leading to the firing of the engineers. They subsequently sued BART and were aided in their suit by the IEEE, which contended that they were performing their ethical duties as engineers in trying to protect the safety of the public that would use BART. Eventually, the engineers were forced to settle the case out of court for only a fraction of the damages that they were seeking.

SECTIONS

OBJECTIVES

After reading this chapter, you will be able to:

- Discuss the responsibilities and rights that engineers have.
- Determine what whistleblowing is and when it is appropriate to blow the whistle.
- Examine issues surrounding the engineer's duty to the environment.
- See how computers have been used unethically.

There are many rights and responsibilities that engineers must exercise in the course of their professional careers. Often, these rights and responsibilities overlap. For example, the BART engineers had a responsibility to the public to see that the BART system was safe and the right to have their concerns taken seriously by management without risking their jobs. Unfortunately, in this case, their rights and responsibilities were not respected by BART. In this chapter, we will take a closer look at these and other rights and responsibilities of engineers.

6.1 INTRODUCTION

The codes of ethics of the professional engineering societies spell out, sometimes in great detail, the responsibilities entailed in being an engineer. However, the codes don't discuss any of the professional rights that engineers should enjoy. There is often a great deal of overlap between these rights and responsibilities. As we saw in the BART case described at the beginning of this chapter, an engineer has a duty to protect the public, by blowing the whistle if necessary, when he perceives that something improper is being done in his organization. The engineer has a right to do this even if his employer feels that it is bad for the organization.

In this chapter, we will discuss the engineer's responsibilities in more detail and also look at the rights of engineers, especially with regard to issues of conscience and conflicts with the rights of employers or clients.

6.2 PROFESSIONAL RESPONSIBILITIES

We will begin our discussion of professional rights and responsibilities by first looking more closely at a few of the important responsibilities that engineers have.

Confidentiality and Proprietary Information

A hallmark of the professions is the requirement that the professional keep certain information of the client secret or confidential. Confidentiality is mentioned in most engineering codes of ethics. This is a well-established principle in professions such as medicine, where the patient's medical information must be kept confidential, and in law, where "attorney–client privilege" is a well-established doctrine. This requirement applies equally well to engineers, who have an obligation to keep proprietary information of their employer or client confidential.

Why must some engineering information be kept confidential? Most information about how a business is run, its products and its suppliers, directly affects the company's ability to compete in the marketplace. Such information can be used by a competitor to gain advantage or to catch up. Thus, it is in the company's (and the employee's) best interest to keep such information confidential to the extent possible.

What types of information should be kept confidential? Some of these types are very obvious, including test results and data, information about upcoming unreleased products, and designs or formulas for products. Other information that should be kept confidential is not as obvious, including business information such as the number of employees working on a project, the identity of suppliers, marketing strategies, production costs, and production yields. Most companies have strict policies regarding the disclosure of business information and require that all employees sign them. Frequently, internal company communications will be labeled as "proprietary." Engineers working for a client are frequently required to sign a nondisclosure agreement. Of course, those engineers working for the government, especially in the defense industry, have even more stringent requirements about secrecy placed on them and may even require a security clearance granted after investigation by a governmental security agency before being able to work.

It seems fairly straightforward for engineers to keep information confidential, since it is usually obvious what should be kept confidential and from whom it should be kept. However, as in many of the topics that we discuss in the context of engineering ethics, there are gray areas that must be considered. For example, a common problem is the question of how long confidentiality extends after an engineer leaves employment with a company. Legally, an engineer is required to keep information confidential even after she has moved to a new employer in the same technical area. In practice, doing so can be difficult. Even if no specific information is divulged to a new employer, an engineer takes with her a great deal of knowledge of what works, what materials to choose, and what components not to choose. This information might be considered proprietary by her former employer. However, when going to a new job, an engineer can't be expected to forget all of the knowledge already gained during years of professional experience.

The courts have considered this issue and have attempted to strike a balance between the competing needs and rights of the individual and the company. Individuals have the right to seek career advancement wherever they choose, even for a competitor of his current employer. Companies have the right to keep information away from their competitors. The burden of ensuring that both of these competing interests are recognized and maintained lies with the individual engineer.

Conflict of Interest

Avoiding conflict of interest is important in any profession, and engineering is no exception. A conflict of interest arises when an interest, if pursued, could keep a professional from meeting one of his obligations [Martin and Schinzinger, 1989]. For example, a civil engineer working for a state department of highways might have a financial interest in a company, that has a bid on a construction project. If that engineer has some responsibility for determining which company's bid to accept, then there is a clear conflict of interest. Pursuing his financial interest in the company might lead him not to objectively and faithfully discharge his professional duties to his employer, the highway department. The engineering codes are very clear on the need to avoid conflicts of interest like this one.

There are three types of conflicts of interest that we will consider [Harris, Pritchard, and Rabins, 1995]. First, there are actual conflicts of interest, such as the one described in the previous paragraph, which compromise objective engineering judgement. There are also potential conflicts of interest, which threaten to easily become actual conflicts of interest. For example, an engineer might find herself becoming friends with a supplier for her company. Although this situation doesn't necessarily constitute a conflict, there is the potential that the engineer's judgement might become conflicted by the needs to maintain the friendship. Finally, there are situations in which there is the appearance of a conflict of interest. This might occur when an engineer is paid based on a percentage of the cost of the design. There is clearly no incentive to cut costs in this situation, and it may appear that the engineer is making the design more expensive simply to generate a larger fee. Even cases where there is only an appearance of a conflict of interest can be significant, because the distrust that comes from this situation compromises the engineer's ability to do this work and future work and calls into question the engineer's judgement.

A good way to avoid conflicts of interest is to follow the guidance of company policy. In the absence of such a policy, asking a coworker or your manager will give you a second opinion and will make it clear that you aren't trying to hide something. In the absence of either of these options, it is best to examine your motives and use some of the ethical problem-solving techniques given in Chapter 4. Finally, you can look to the statements in the professional ethics codes that uniformly forbid conflicts of interest. Some of the codes have very explicit statements that can help determine whether or not your situation is a conflict of interest.

Environmental Ethics

One of the most important political issues of the late 20th century has been environmental protection and the rise of the environmental movement. This movement has sought to control the introduction of toxic and unnatural substances into the environment, to protect the integrity of the biosphere, and to ensure a healthy environment for humans. Engineers are responsible in part for the creation of the technology that has led to damage of the environment and are also working to find solutions to the problems caused by modern technology. The environmental movement has led to an increased awareness among engineers that they have a responsibility to use their knowledge and skills to help protect the environment. This duty is even spelled out in the code of ethics of the IEEE.

As concern about the environment has grown, ethicists have turned their attention to the ethical dimensions of environmentalism. In the late 1960s, an area of study called "environmental ethics" was formulated, seeking to explore the ethical roots of the environmental movement and to understand what ethics tells us about our responsibility to the environment.

Fundamental to discussing ethical issues in environmentalism is a determination of the moral standing of the environment. Our Western ethical tradition is anthropocentric, meaning that only human beings have moral standing. Animals and plants are important only in respect to their usefulness to humans. This type of thinking is often evident even within the environmental movement when a case is sometimes made for the protection of rare plants based on their potential for providing new medicines. If animals, trees, and other components of the environment have no moral standing, then we have no ethical obligations towards them beyond maintaining their usefulness to humans. There are, however, other ways to view the moral standing of the environment.

One way to explore the environment's moral status is to try to answer some questions regarding the place of humans in our environment. Do we belong to nature, or does nature belong to us? If animals can suffer and feel pain like humans, should they have moral standing? If animals have moral standing, how far does this moral standing then extend to other life forms, such as trees? Clearly, these questions are not easily answered, and not everyone will come to the same conclusions. However, there are significant numbers of people who feel that the environment, and specifically animals and plants, do have standing beyond their usefulness to humans. In one form, this view holds that humans are just one component of the environment and that all components have equal standing. For those who hold this view, it is an utmost duty of everyone to do what is required to maintain a healthy biosphere for its own sake.

Regardless of the goal (i.e., either protecting human health or protecting the overall health of the biosphere for its own sake), there are multiple approaches that can be taken to resolving environmental problems. Interestingly, these approaches mirror the general approaches to ethical problem solving discussed in Chapter 4. The first approach is sometimes referred to as the "cost-oblivious approach" [Martin and Schinzinger, 1989]. In this approach, cost is not taken into account, but rather the environment is made as clean as possible. No level of environmental degradation is seen as acceptable. This approach bears a striking resemblance to rights and duty ethics, as discussed in Chapter 3. There are obvious problems with this approach. It is difficult to uphold, especially in a modern urbanized society. It is also very difficult to enforce, since the definition of "as clean as possible" is hard to agree on, and being oblivious to cost isn't practical in any realistic situation, in which there are not infinite resources to apply to a problem.

A second approach is based on cost–benefit analysis, which is derived from utilitarianism. Here, the problem is analyzed in terms of the benefits derived by reducing the pollution—improvements in human health, for example—and the costs required to solve the problem. The costs and benefits are weighed to determine the optimum combination. In this approach, the goal is not to achieve a completely clean environment, but rather to achieve an economically beneficial balance of pollution with health or environmental considerations.

There are problems associated with the cost–benefit approach. First, there is an implicit assumption in cost–benefit analysis that cost is an important issue. But what is the true cost of a human life or the loss of a species or a scenic view? These values are difficult, if not impossible, to determine. Second, it is difficult to accurately assess costs and benefits, and much guesswork must go into these calculations. Third, this approach doesn't necessarily take into account who shoulders the costs and who gets the benefits. This is frequently a problem with the siting of landfills and other waste dumps. The cheapest land is in economically disadvantaged areas, where people don't necessarily have the political clout, education, or money required to successfully oppose a landfill in their neighborhood. Although dumps have to go somewhere, there should be some attempt to share the costs as well as share the benefits of an environmentally questionable project. Finally, cost–benefit analysis doesn't necessarily take morality or ethics into account. The only considerations are costs and benefits, with no room for a discussion of whether what is being done is right or not.

Given the complexity of these issues, what then are the responsibilities of the engineer to the environment? When looking at the environmental aspects of her work, an engineer can appeal to both professional and personal ethics to make a decision. Of course, the minimal requirement is that the engineer must follow the applicable federal, state, and municipal laws and regulations.

Professional codes of ethics tell us to hold the safety of people and the environment to be of paramount importance. So clearly, engineers have a responsibility to ensure that their work is conducted in the most environmentally safe manner possible. This is true certainly from the perspective of human health, but for those who feel that the environment has moral standing of its own, the responsibility to protect the environment is clear. Often, this responsibility must be balanced somewhat by consideration of the economic well-being of our employer, our family, and our community.

Our personal ethics can also be used to determine the best course when we are confronted with an environmental problem. Most of us have very strong beliefs about the need to protect the environment. Although these beliefs may come into conflict with our employer's desires, we have the right and duty to strongly express our views on what is acceptable. As we will see later in this chapter, as professionals, engineers have the right to express their opinions on moral issues such as the environment. An engineer should not be compelled by his employer to work on a project that he finds ethically troubling, including projects with severe environmental impacts.

In trying to decide what the most environmentally acceptable course of action is, it is also important to remember that a basic tenet of professional engineering codes of ethics states that an engineer should not make decisions in areas in which he isn't competent. For many environmental issues, engineers aren't competent to make decisions, but should instead seek the counsel of others—such as biologists, public-health experts, and physicians—who have the knowledge to help analyze and understand the possible environmental consequences of a project.

6.3 COMPUTER ETHICS

Computers have rapidly become a ubiquitous tool in engineering and business. There are ways in which computers have brought benefits to society. Unfortunately, there are also numerous ways in which computers have been misused, leading to serious ethical issues. The engineer's roles as designer, manager, and user of computers bring with them a responsibility to help foster the ethical use of computers.

We will see that the ethical issues associated with computers are really just variations on the issues that we have already dealt with in this book. For example, many ethical problems associated with computer use relate to unauthorized use of information stored on computer databases and are thus related to the issues of confidentiality and proprietary information discussed in the previous section. The ethical problem-solving techniques developed previously in this book will be equally applicable to computer ethics issues.

There are three broad categories of computer ethics problems: those for which the computer is the instrument of the unethical act, such as the use of a computer to defraud a bank; those for which the computer is the object of the act, as when computer software is stolen and installed on one's own computer or when information is accessed from someone else's computer; and those problems associated with the autonomous nature of computers [Martin and Schinzinger, 1989].

Computers as the Instrument of Unethical Behavior

Our discussion of computer ethics will start with an examination of ways in which computers are used as the instruments of unethical behavior. Many of these uses are merely extensions to computers of other types of unethical acts. For example, computers can be used to more efficiently steal money from a bank. A more traditional bank-robbery method is to put on a mask, hand a note to a bank teller, show your gun, and walk away with some cash. Computers can be used to make bank robbery easier to perform and harder to trace. The robber simply sits at a computer terminal—perhaps the modern equivalent of a mask—invades the bank's computer system, and directs that some of the bank's assets be placed in a location accessible to him. Using a computer, a criminal can also make it difficult for the theft to be detected and traced.

It is clear that from an ethical standpoint, there is no difference between a bank robbery perpetrated in person or one perpetrated via a computer, although generally the amounts taken in a computer crime far exceed those taken in an armed robbery. The difference between these two types of robbery is that the use of the computer makes the crime impersonal. The criminal never comes face to face with the victim. In addition, the use of the computer makes it easier to steal from a wide variety of people. Computers can be used to steal from an employer, outsiders can get into a system and steal from an institution such as a bank, or a company can use the computer to steal from its clients and customers. In these cases, the computer has only made the theft easier to perpetrate, but does not alter the ethical issues involved. Unfortunately, the technology to detect and prevent this type of crime greatly lags behind the computer technology available to commit it. Those seeking to limit computer crime are always playing a catch-up game.

Another instrumental area of computer ethics involves privacy. It is widely held that certain information is private and cannot be divulged without consent. This includes information about individuals as well as corporate information. Computers did not create the issues involved in privacy, but they certainly have exacerbated them. Computers make privacy more difficult to protect, since large amounts of data on individuals and corporations are centrally stored on computers where an increasing number of individuals can access it. Before we look at the ways that privacy can be abused by the use of computers, we will discuss the issues surrounding privacy and see what the ethical standing of privacy is.

By privacy, we mean the basic right of an individual to control access to and use of information about himself [Martin and Schinzinger, 1989]. Why is privacy an ethical issue? Invasions of privacy can be harmful to an individual in two ways. First, the leaking of private information can lead to an individual's being harassed or blackmailed. In its most simple form, this harassment may come in the form of repeated phone calls from telemarketers who have obtained information about an individual's spending habits. The harassment might also come in the form of subtle teasing or bothering from a coworker who has gained personal knowledge of the individual. Clearly, individuals have the right not to be subjected to this type of harassment. Second, personal information can also be considered personal property. As such, any unauthorized use of this information is theft. This same principle applies to proprietary information of a corporation.

How do computers increase the problems with privacy protection? This phenomenon is most easily seen by looking at the old system of record keeping. For example, medical records of individuals were at one time kept only on paper and generally resided with the individual's physician and in hospitals where a patient had been treated. Access to these records by researchers, insurance companies, or other healthcare providers was a somewhat laborious process involving searching through storage for the appropriate files, copying them, and sending them through the mail. Unauthorized use of this information involved breaking into the office where the files were kept and stealing them or, for those who had access to the files, surreptitiously removing the files. Both of these acts involved a substantial risk of being caught and prosecuted. Generally, these records have now been computerized. Although computerization makes the retrieval of files much easier for those with legitimate needs and reduces the space required to store the files, it also makes the unauthorized use of this information by others easier.

Computers as the Object of Unethical Acts

Ethical issues also arise when computers are the object of an unethical act. This act is popularly referred to as "hacking" and has been widely reported in the newspapers and in popular culture, sometimes with the "hacker" being portrayed as heroic. Hacking comes in many forms: gaining unauthorized access to a database, implanting false information in a database or altering existing information, and disseminating viruses over the Internet.

These activities are by no means limited to highly trained computer specialists. Many hackers are bored teenagers seeking a challenge. Computer hacking is clearly ethically troublesome. As mentioned before, accessing private information violates the privacy rights of individuals or corporations, even if the hacker keeps this information to himself. In extreme cases, hackers have accessed secret military information, which has obvious implications for national security. Altering information in a database, even information about yourself, is also ethically troubling, especially if the alteration has the intent of engaging in a fraud.

The issuance of computer viruses is also unethical. These viruses frequently destroy data stored on computers. In extreme cases, this act could lead to deaths when hospital records or equipment are compromised, to financial ruin for individuals whose records are wiped out, or even to the loss of millions of dollars for corporations, individuals, and taxpayers, as completed work must be redone after being destroyed by a virus.

Oftentimes, hackers are not being malicious, but are simply trying to "push the envelope" and see what they and their computers are capable of. Nevertheless, hacking is an unethical use of computers.

Autonomous Computers

Other ethical concerns arise because of the increasingly autonomous nature of computers. Autonomy refers to the ability of a computer to make decisions without the intervention of humans. Some of the negative implications of this autonomy are chillingly spelled out in *2001: A Space Odyssey*, by Arthur C. Clarke, in which an autonomous computer responsible for running a spaceship headed for Jupiter begins to turn against the humans it was designed to work for. Certainly, there are applications for which autonomy is valuable. For example, manufacturing processes that require monitoring and control at frequent intervals can greatly benefit from autonomous computers. In this case, the autonomy of the computer has very little impact beyond the interests of the manufacturer.

Other autonomous computer applications are not so benign. For example, by the 1980s, computers were widely used to automate trading on the major U.S. stock exchanges. Some brokerages and institutional investors utilized computers that were programmed to sell stocks automatically under certain conditions, among them when prices drop sharply. This type of programming creates an unstable situation. As prices drop, computers automatically start selling stocks, further depressing the prices, causing other computers to sell, and so on until there is a major market crash.

This scenario actually occurred on October 19, 1987, when the Dow Jones Industrial Average (a widely used market-price indicator) dropped by 508 points, a 22.6% drop in the overall value of the market. Interestingly, during the famous October 1929 stock market crash that launched the Great Depression, the percent drop in overall market value was only half of this amount. The 1987 crash was widely attributed to automated computer trading. Federal regulations have since been implemented to help prevent a recurrence of this problem.

Autonomy of computer systems has also been called into question with regard to military weapons. Many weapons systems rely heavily on computer sensors and computer controls. Due to the speed with which events can happen on a modern battlefield, it would seem valuable to have weapons that can operate autonomously. However, weapons systems operating without human intervention can suffer from the instability problems described with regard to the financial markets. For example, a malfunctioning sensor might lead a computer to think that an enemy has increased its military activity in a certain area. This would lead to an increased readiness on our part, followed by increased activity by the enemy, etc. This unstable situation could lead to a conflict and the loss of life when really there was nothing happening [Rauschenbakh, 1988]. This problem is especially concerning due to the implications for loss of human life. It is clear from this example that although autonomous computers can greatly increase productivity and efficiency in many areas, ultimately there must be some human control in order to prevent disasters.

Computer Codes of Ethics

To aid with decision making regarding these and other computer-related ethics issues, many organizations have developed codes of ethics for computer use. The discussion in Chapter 2 regarding the purposes of codes and the way in which codes of ethics function is equally true for codes related to computer use. They are guidelines for the ethical use of computing resources, but should not be used as a substitute for sound moral reasoning and judgement. They do, however, provide some guidance in the proper use of computer equipment.

6.4 PROFESSIONAL RIGHTS

We have seen how the professional status of engineering confers many responsibilities on the engineer. Engineers also have rights that go along with these responsibilities. Not all of these rights come about due to the professional status of engineering. There are rights that individuals have regardless of professional status, including the right to privacy, the right to participate in activities of one's own choosing outside of work, the right to reasonably object to company policies without fear of retribution, and the right to due process.

The most fundamental right of an engineer is the right of professional conscience [Martin and Schinzinger, 1989]. This involves the right to exercise professional judgement in discharging one's duties and to exercise this judgement in an ethical manner. This right is basic to an engineer's professional practice. However, it is no surprise that this right is not always easy for an employer to understand.

The right of professional conscience can have many aspects. For example, one of these aspects might be referred to as the "Right of Conscientious Refusal" [Martin and Schinzinger, 1989]. This is the right to refuse to engage in unethical behavior. Put quite simply, no employer can ask or pressure an employee into doing something that she considers unethical and unacceptable. Although this issue is very clear in cases for which an engineer is asked to falsify a test result or fudge on the safety of a product, it is less clear in cases for which the engineer refuses an assignment based on an ethical principle that is not shared by everyone. For example, an engineer ought to be allowed to refuse to work on defense projects or environmentally hazardous work if his conscience says that such work is immoral. Employers should be reasonably accommodating of that person's request. We will amplify this point with regard to defense work in the next section.

Engineers and the Defense Industry

One of the largest employers of engineers worldwide is the defense industry. This is by no means a modern trend; throughout history, many innovations in engineering and science have come about as the result of the development of weapons. Since fundamentally, weapons are designed for one purpose—to kill human beings—it seems important to look at this type of engineering work in the context of engineering ethics and the rights of engineers.

An engineer may choose either to work or not to work in defense-related industries and be ethically justified in either position. Many reasonable engineering professionals feel that ethically, they cannot work on designs that will ultimately be used to kill other humans. Their remoteness from the killing doesn't change this feeling. Even though they won't push the button or may never actually see the victims of the use of the weapon, they still find it morally unacceptable to work on such systems.

On the other hand, equally morally responsible engineers find this type of work ethically acceptable. They reason that the defense of our nation or other nations from aggression is a legitimate function of our government and is an honorable goal for engineers to contribute to. Both of these positions can be justified by reference to the moral theories and problem solving techniques that were presented previously in this book.

Even if an engineer finds defense work ethically acceptable, there might be uses of these weapons or certain projects that he considers questionable. For example, is it acceptable to work on weapons systems that will only be sold to other nations? Is the use of weapons to guarantee our "national interests," such as maintaining a steady supply of foreign oil, an acceptable defense project?

Given the issues that surround defense work, what is an engineer to do when asked to work on a weapons project he considers questionable? As with many of the ethical dilemmas that we have discussed in this book, there is no simple solution, but rather the answer must be determined by each individual after examination of his values and personal feelings about the ethics of defense work. It is important to avoid working on any project that you deem unethical, even if it might lead to a career advancement, or even if it is a temporary job. (This principle also holds true for projects that you feel are unsafe, bad for the environment, etc.) It can be argued that weapons work is the most important type of engineering, given its consequences for mankind. Because of the implications to human life, this type of engineering requires an even more stringent examination of ethical issues to ensure responsible participation.

6.5 WHISTLEBLOWING

There has been increased attention paid in the last 30 years to whistleblowing, both in government and in private industry. Whistleblowing is the act by an employee of informing the public or higher management of unethical or illegal behavior by an employer or supervisor. There are frequent newspaper reports of cases in which an employee of a company has gone to the media with allegations of wrongdoing by his employer or in which a government employee has disclosed waste or fraud. In this section, we will examine the ethical aspects of whistleblowing and discuss when it is appropriate and when it isn't appropriate. We will also look at what corporations and government agencies can do to lessen the need for employees to take this drastic action.

Whistleblowing is included in this chapter on rights and responsibilities because it straddles the line between both. According to the codes of ethics of the professional engineering societies, engineers have a duty to protect the health and safety of the public, so in many cases, an engineer is compelled to blow the whistle on acts or projects that harm these values. Engineers also have the professional right to disclose wrongdoing within their organizations and expect to see appropriate action taken.

Types of Whistleblowing

We will start our discussion of whistleblowing by looking at the different forms that whistleblowing takes. A distinction is often made between internal and external whistleblowing. Internal whistleblowing occurs when an employee goes over the head of an immediate supervisor to report a problem to a higher level of management. Or, all levels of management are bypassed, and the employee goes directly to the president of the company or the board of directors. However it is done, the whistleblowing is kept within the company or organization. External whistleblowing occurs when the employee goes outside the company and reports wrongdoing to newspapers or law-enforcement authorities. Either type of whistleblowing is likely to be perceived as disloyalty. However, keeping it within the company is often seen as less serious than going outside of the company.

There is also a distinction between acknowledged and anonymous whistleblowing. Anonymous whistleblowing occurs when the employee who is blowing the whistle refuses to divulge his name when making accusations. These accusations might take the form of anonymous memos to upper management (as in the BART case discussed later) or of anonymous phone calls to the police. The employee might also talk to the news media but refuse to let her name be used as the source of the allegations of wrongdoing. Acknowledged whistleblowing, on the other hand, occurs when the employee puts his name behind the accusations and is willing to withstand the scrutiny brought on by his accusations.

Whistleblowing can be very bad from a corporation's point of view because it can lead to distrust, disharmony, and an inability of employees to work together. The situation can be illustrated by an analogy with sports. If the type of whistleblowing we are discussing here was performed during a game, it would not be the referees who stopped play because of a violation of the rules. Rather, it would be one of your own teammates who stopped the game and assessed a penalty on your own team. In sports, this type of whistleblowing would seem like an act of extreme disloyalty, although perhaps it is the "gentlemanly" thing to do. Similarly, in business, whistleblowing is perceived as an act of extreme disloyalty to the company and to coworkers.

When Should Whistleblowing Be Attempted?

During the course of your professional life, you might come across a few cases of wrongdoing. How do you know when you should blow the whistle? We will start to answer this question by first looking at when you *may* blow the whistle and then looking at when you *should* blow the whistle. Whistleblowing should only be attempted if the following four conditions are met [Harris, Pritchard, and Rabins, 1995]:

1. *Need.* There must be a clear and important harm that can be avoided by blowing the whistle. In deciding whether to go public, the employee needs to have a sense of proportion. You don't need to blow the whistle about everything, just the important things. Of course, if there is a pattern of many small things that are going on, this can add up to a major and important matter requiring that the whistle be blown. For example, if an accident occurs at your company, resulting in a spill of a small quantity of a toxic compound into a nearby waterway that is immediately cleaned up, this incident probably does not merit notifying outside authorities. However, if this type of event happens repeatedly and no action is taken to rectify the problem despite repeated attempts by employees to get the problem fixed, then perhaps this situation is serious enough to warrant the extreme measure of whistleblowing.
2. *Proximity.* The whistleblower must be in a very clear position to report on the problem. Hearsay is not adequate. Firsthand knowledge is essential to making an effective case about wrongdoing. This point also implies that the whistleblower must have enough expertise in the area to make a realistic assessment of the situation. This condition stems from the clauses in several codes of ethics that mandate that an engineer not undertake work in areas outside her expertise. This principle applies equally well to making assessments about whether wrongdoing is taking place.
3. *Capability.* The whistleblower must have a reasonable chance of success in stopping the harmful activity. You are not obligated to risk your career and the

financial security of your family if you can't see the case through to completion or you don't feel that you have access to the proper channels to ensure that the situation is resolved.

4. *Last resort.* Whistleblowing should be attempted only if there is no one else more capable or more proximate to blow the whistle and if you feel that all other lines of action within the context of the organization have been explored and shut off.

These four conditions tell us when whistleblowing is morally acceptable. But when is an engineer morally obligated to blow the whistle? There may be situations in which you are aware of wrongdoing and the four conditions discussed above have been met. In this case, the whistle *may* be blown if you feel that the matter is sufficiently important. You are only *obligated* to blow the whistle when there is great imminent danger of harm to someone if the activity continues and the four conditions have been met. A great deal of introspection and reflection is required before whistleblowing is undertaken.

It is important for the whistleblower to understand his motives before undertaking this step. It is acceptable to blow the whistle to protect the public interest, but not to exact revenge upon fellow employees, supervisors, or your company. Nor is it acceptable to blow the whistle in the hopes of future gains through book contracts and speaking tours.

Preventing Whistleblowing

So far, our discussion of whistleblowing has focused on the employee who finds herself in a situation where she feels that something must be done. We should also look at whistleblowing from the employer's point of view. As an employer, I should seek to minimize the need for employees to blow the whistle within my organization. Clearly, any time that information about wrongdoing becomes public, it is harmful to the organization's image and will negatively affect the future prospects of the company. How, then, do I stop this type of damage?

In answering this question, we must acknowledge that it is probably impossible to eliminate all wrongdoing in a corporation or government agency. Even organizations with a very strong ethical culture will have employees who, from time to time, succumb to the temptation to do something wrong. A typical corporate approach to stemming whistleblowing and the resulting bad publicity is to fire whistleblowers and to intimidate others that might seem likely to blow the whistle. This type of approach is both ineffective and ethically unacceptable. No one should be made to feel bad about trying to stop ethically questionable activities.

There are four ways in which to solve the whistleblowing problem within a corporation. First, there must be a strong corporate ethics culture. This should include a clear commitment to ethical behavior, starting at the highest levels of management, and mandatory ethics training for all employees. All managers must set the tone for the ethical behavior of their employees. Second, there should be clear lines of communication within the corporation. This openness gives an employee who feels that there is something that must be fixed a clear path to air his concerns. Third, all employees must have meaningful access to high-level managers in order to bring their concerns forward. This access must come with a guarantee that there will be no retaliation. Rather, employees willing to come forward should be rewarded for their commitment to fostering the ethical behavior of the company. Finally, there should be willingness on the part of management to admit mistakes, publicly if necessary. This attitude will set the stage for ethical behavior by all employees

APPLICATION: CASES

The BART Case

The cities surrounding San Francisco Bay form one of the largest metropolitan areas in the United States. Due to the geographical limits imposed by the bay, much of the commuting that takes place in this area must be across just a few bridges. The Bay Area Rapid Transit system (BART) had its genesis in late 1947 when a joint Army–Navy review board recommended the construction of a tunnel underneath San Francisco Bay for high-speed train service between San Francisco and Oakland [Friedlander, 1972]. The California state legislature then formed the San Francisco Bay Area Rapid Transit Commission, which was to study the transportation needs of the Bay area and make recommendations to the legislature. This effort culminated in the formation of the Bay Area Rapid Transit district in 1957. By 1962, this group had done a preliminary design of a rapid train system, including a transbay tube, and had laid the groundwork for fund-raising for the project. In 1962, a bond issue to fund the project was approved by the voters and the project was begun.

As envisioned, BART was to be a high-tech rail system serving many of the outlying communities along San Francisco Bay. There were three distinct engineering issues involved in BART: the design and construction of railbeds, tunnels, bridges, etc.; the design and manufacture of the railcars; and design and implementation of a system for controlling the trains. The control system will be the focus of our discussion.

BART was to incorporate much new technology, including fully automated control systems. The trains would have "attendants," but would not be under direct control by humans. In many respects, BART was an experiment on a very large scale. None of the control technologies that were to be used had been previously tested in a commuter rail system. Of course, any innovative engineering design is like this and has components that have not been previously tested.

The Automatic Train-Control (ATC) system was an innovative method for controlling train speed and access to stations. In most urban mass transit systems, this function is performed by human drivers reading trackside signals and receiving instructions via radio from dispatchers. Instead, BART relied on a series of onboard sensors that determined the train's position and the location of other trains. Speeds on the track were automatically maintained by monitoring the location of the train and detecting allowed speed information. One of the unique and problematic features of the system was that there were no fail-safe methods of train control [Friedlander, 1972]. Rather, all control was based on redundancy. This distinction is very important. "*Fail safe*" implies that if there is a failure, the system will revert to a safe state. In the case of BART, this would mean that a failure would cause the trains to stop. Redundancy, on the other hand, relies on switching failed components or systems to backups in order to keep the trains running.

There are two distinct phases of this type of engineering project, construction and operation, each requiring different skills. For this reason, early on, BART decided to keep its own staff relatively small and subcontract most of the design and construction work. This way, there wouldn't be the need to lay off hundreds of workers during the transition from construction to operation [Anderson, 1980]. This system also encouraged the engineers who worked for BART not only to oversee the design and construction of the system, but also to learn the skills required to run and manage this complex transportation system. Contracts for design and construction of the railroad infrastructure were awarded to a consortium of large engineering firms known as Parsons, Brinkerhoff, Tudor, and Bechtel (PBTB). PBTB began construction on the system in January of 1967. The transbay tube was started in November of that year. Also in 1967, a contract was awarded to Westinghouse to design and build the ATC. In 1969, Rohr industries was awarded a contract to supply 250 railroad cars.

A little bit should be said about the management structure at BART. By design, BART was organized with a very open management structure. Employees were given great freedom to define what their jobs entailed and to work independently and were encouraged to take any concerns that they had to management

Unfortunately, there was also a very diffuse and unclear chain of command that made it difficult for employees to take their concerns to the right person [Anderson, 1980].

The key players in this case were three BART engineers working on various aspects of the ATC: Roger Hjortsvang, Robert Bruder, and Max Blankenzee. The first to be employed by BART was Hjortsvang. As part of his duties for BART, Hjortsvang spent 10 months in 1969–70 in Pittsburgh at the Westinghouse plant working with the engineers who were designing the ATC. During this time, he became concerned about the lack of testing of some of the components of the ATC and also about the lack of oversight of Westinghouse by BART. After returning to San Francisco, Hjortsvang began raising some of these concerns with his management.

Soon after Hjortsvang returned from Pittsburgh, Bruder joined BART, working in a different group than Hjortsvang. He also became concerned about the Westinghouse test procedures and about the testing schedule, but was unable to get his concerns addressed by BART management. Both Hjortsvang and Bruder were told that BART management was satisfied with the test procedures Westinghouse was employing. Management felt that Westinghouse had been awarded the contract because of its experience and engineering skills and should be trusted to deliver what was promised.

Around this time, both engineers also became concerned about the documentation that Westinghouse was providing. Would the documentation be sufficient for BART engineers to understand how the system worked? Would they be able to repair it or modify it once the system was delivered and Westinghouse was out of the picture? Being unable to get satisfaction, Hjortsvang and Bruder dropped the matter. It is important to note that the concerns here were not just about testing, *per se,* but also about the effect that untested components might have on the safety and reliability of BART.

Blankenzee then joined BART and worked at the same location as Hjortsvang. Before joining BART, Blankenzee had worked for Westinghouse on the BART project, and so he knew about how Westinghouse was approaching its work. He too was concerned about the testing and documentation of the ATC. When Blankenzee joined BART, it rekindled Hortsvang's and Bruder's interest in these problems. To attempt to resolve these concerns, Hortsvang wrote an unsigned memo in November of 1971 to several levels of BART management that summarized the problems he perceived. Distribution of an anonymous memo was, of course, viewed with suspicion by management.

In January 1972, the three engineers contacted members of the BART board of directors, indicating that their concerns were not being taken seriously by lower management. This action was in direct conflict with the general manager of BART, whose policy was to allow only himself and a few others to deal directly with the board [Anderson, 1980]. As defined previously in this chapter, this action by the engineers constituted "internal whistleblowing." The engineers also consulted with an outside engineering consultant, Edward Burfine, who evaluated the ATC on his own and came to conclusions similar to those of the three engineers.

One of the members of the board of directors, Dan Helix, spoke with the engineers and appeared to take them seriously. Helix took the engineer's memos and the report of the consultant and distributed them to other members of the board. Unfortunately, he also released them to a local newspaper, a surprising act of external whistleblowing by a member of the board of directors. Naturally, BART management was upset by this action and tried to locate the source of this information. The three engineers initially lied about their involvement. They later agreed to take their concerns directly to the board, thus revealing themselves as the source of the leaks. The board was skeptical of the importance of their concerns. Once the matter was in the open, the engineers' positions within BART became tenuous.

On March 2 and 3, 1972, all three engineers were offered the choice of resignation or firing. They all refused to resign and were dismissed on the grounds of insubordination, lying to their superiors (they had denied being the source of the leaks), and failing to follow organizational procedures. They all suffered as a result of their dismissal. None was able to find work for a number of months, and all suffered financial and emotional problems as a result. They sued BART for $875,000, but were forced to settle out of court, since it was likely that their lying to superiors would be very detrimental to the case. Each received just $25,000 [Anderson, 1980].

As the legal proceedings were taking place, the IEEE attempted to assist the three engineers by filing an *amicus curiae* (friend of the court) brief in their support. The IEEE asserted that each of the engineers had a professional duty to keep the safety of the public paramount and that their actions were therefore justified. Based on the IEEE code of ethics, the brief stated that engineers must "notify the proper authority of any observed conditions which endanger public safety and health." The brief interpreted this statement to mean that in the case of public employment, the proper authority is the public itself [Anderson, 1980]. This was perhaps the first time that a national engineering professional society had intervened in a legal proceeding on behalf of engineers who had apparently been fulfilling their duties according to a professional code of ethics.

Safety concerns continued to mount as BART was put into operation. For example, on October 2, 1972, less than a month after BART was put into revenue service, a BART train overshot the station at Fremont, California and crashed into a sand embankment. There were no fatalities, but five persons were injured. The accident was attributed to a malfunction of a crystal oscillator, part of the ATC, which controlled the speed commands for the train. Subsequent to this accident, there were several investigations and reports on the operation of BART. These revealed that there had been other problems and malfunctions in the system. Trains had often been allowed too close to each other; sometimes a track was indicated to be occupied when it wasn't and was indicated not to be occupied when it was. The safety concerns of the three engineers seemed to be borne out by the early operation of the system [Friedlander, 1972, 1973].

Ultimately, the ATC was improved and the bugs worked out. In the years since, BART has accumulated an excellent safety record and has served as the model for other high-tech mass transit systems around the country.

The Goodrich A7-D Brake Case

This case is one that is very often used as an example in engineering ethics texts, especially to study whistleblowing. In studying this case, it is important to keep in mind that much of the information presented here is derived from the writing of the whistleblower. An individual who is deeply embroiled in a controversial situation such as this one will have different insights and viewpoints on the situation than will management or other workers. Little is publicly known about what Goodrich management thought about this case.

In the 1960s, the B. F. Goodrich corporation was a major defense contractor. One of their main defense related industries was the production of brakes and wheels for military aircraft. This activity was located in Troy, Ohio. Goodrich had developed a new and innovative design: a four-rotor brake that would be considerably lighter than the more traditional five-rotor designs. Any reduction in weight is very attractive in aircraft design, since it allows for an increase in payload weight with no decrease in performance.

In June of 1967, Goodrich was awarded the contract to supply the brakes for the A7–D by LTV, the prime contractor for the airplane. The qualifying of this new design was on a very tight schedule imposed by the Air Force. The new brake had to be ready for flight testing by June of 1968, leaving only one year to test and qualify the design. To qualify the design for the flight test, Goodrich had to demonstrate that it performed well in a series of tests specified by the Air Force.

After the design had been completed, John Warren, the design engineer, handed the project over to Searle Lawson, who was just out of engineering school, to perform the testing of the brakes. Warren moved on to other projects within the corporation. Lawson's first task was to test various potential brake-lining materials to see which ones would work best in this new design. This test would be followed by the testing of the chosen linings on full-scale prototypes of the brakes. Unfortunately, after six months of testing, Lawson was unable to find any materials that worked adequately. He became convinced that the design itself was flawed and would never perform according to the Air Force's specifications.

Lawson spoke with Warren about these problems. Warren still felt that the brake design was adequate and made several suggestions to Lawson regarding new lining materials that might improve performance. However, none of these suggestions worked and the brakes still failed to pass the initial tests. Lawson then spoke about these problems with Robert Sink, the A7-D project manager at Goodrich. Sink asked Lawson to keep trying some more linings and expressed confidence that the design would work correctly.

In March of 1968, Goodrich began testing the full brake prototypes. After 13 tests, the brake had yet to pass the Air Force's specification for temperature. The only way to get the brakes to pass the test was to set up cooling fans directed at the rotors. Obviously, brakes that required extra cooling would not meet the Air Force's specification. Nevertheless, Sink assured LTV that the brake development was going well.

Kermit Vandivier was a technical writer for Goodrich who was responsible for writing test reports and was assigned to write the report for the new A7-D brakes. This report would be an integral part of the Air Force's decision-making process. Vandivier was not an engineer, but he did have experience in writing up the results of this type of test. In the course of writing the report on the A7-D brake tests, Vandivier became aware that some of the test results had been rigged to meet the Air Force's specifications. Vandivier raised his concerns about the report he was writing, feeling that he couldn't write a report based on falsified data. His attempts to write an accurate report were not allowed by management, and Goodrich submitted a report using the jury-rigged data. Based on this report, the brake was qualified for flight testing.

Vandivier was concerned about the safety of the brake and wondered what his legal responsibility might be. He contacted his attorney, who suggested that he and Lawson might be guilty of conspiracy to commit fraud and advised Vandivier to meet with the U.S. Attorney in Dayton. Upon advice of the U.S. Attorney, both Lawson and Vandivier contacted the FBI.

In July, the Air Force asked Goodrich to supply the raw test data for review. This request led to efforts at Goodrich to control the damage that would ensue when the real nature of the tests became known. Not being satisfied with the report presented to it, the Air Force refused to accept the brake. Knowing that the four-rotor brake was not going to work, Goodrich began an effort to design a five-rotor replacement. Vandivier continued meeting with the FBI and supplied FBI agents with Goodrich documents related to the A7-D brake tests.

Apparently, Lawson had impressed LTV because after the flight testing was over, LTV offered him a job. Lawson accepted and left Goodrich on October 11. With the only other person who really knew about the test procedures gone, Vandivier also decided to resign from Goodrich. In his letter of resignation, he included a series of accusations of wrongdoing against Goodrich regarding the brake tests. Vandivier went to work for the *Troy Daily News*, the local newspaper.

At the *Daily News*, Vandivier told his editor about the situation at Goodrich. From there, the story made its way to Washington, where it came to the attention of Senator William Proxmire, among others. In May of 1969, Proxmire requested that the Government Accounting Office (GAO) review the issue of the qualification testing of the A7-D brakes. The GAO investigation led to an August 1969 Senate hearing chaired by Proxmire. By then, the new five-rotor brake had been tested and qualified for use on the A7-D. At the hearing, Vandivier's concerns and the GAO findings were publicly aired. The GAO report confirmed Vandivier's statements about testing discrepancies, though the report also showed that there was no additional cost to the government in obtaining a working brake and that the brake problems didn't cause any substantial delays in the overall A-7D program.

No official action was taken against Goodrich as a result of this incident, and there does not seem to have been any negative impact on the careers of those at Goodrich involved in the A7-D project. Lawson went on to a successful career at LTV. Vandivier later wrote a chapter of a book and an article in Harper's magazine detailing his version of the story.

The City of Albuquerque vs. Isleta Pueblo Water Case

The city of Albuquerque, New Mexico straddles the Rio Grande and is bounded on the north and south by two Indian pueblos (reservations). Several other pueblos are nearby. According to federal law, Indian tribes are sovereign nations with wide ranging ability to self-regulate and are subject to federal laws and some restrictions imposed by the states. Overall, however, their status is closer to that of an equal of state governments rather than a subordinate.

Isleta Pueblo is located on the Rio Grande, downstream from Albuquerque, and is contiguous to the Albuquerque metropolitan area, which contains approximately 650,000 people. Traditionally, the Pueblo used water directly from the river for drinking during religious ceremonies. In more recent times, this practice has been difficult due to runoff entering the river—storm runoff is directly input to the river—and from treated sewer effluent placed into the river by Albuquerque. Similar effluent is probably discharged into the river by other municipalities farther upstream.

Of great concern to Isleta Pueblo is the concentration of arsenic in the river water. The Albuquerque sewage treatment plant puts water into the Rio Grande that meets all applicable EPA regulations, including the standard for arsenic concentration. Of course, the water placed into the river is not of drinking quality, since it is assumed that any municipality using river water for drinking must treat the water anyway.

Isleta Pueblo has used its sovereign status to try to enforce a stricter water-quality standard for the water discharged by Albuquerque and seeks to bring the water quality to the point where it can be consumed directly from the river. This involves a standard for arsenic discharge that is roughly twice as stringent as the EPA regulations permit. The EPA has sided with the pueblo, citing federal law giving Indian reservations the right to set their own pollution standards. This case is analogous to the situation that might occur if Mexico decided that it wanted stricter regulation of the quality of water in the Rio Grande flowing from the United States south along the Mexican border.

The city of Albuquerque has argued that the pueblo's standards are too strict and are unnecessary, since the concentration of arsenic in the water that the city discharges into the river is lower than what naturally exists in the river upstream from Albuquerque, although this point is under debate. Albuquerque contends that the cost of meeting the standard would be prohibitive, approximately $300 million. The city also argues that the standard is a violation of the First Amendment's prohibition of government-established religion. Albuquerque is currently trying to have the case heard by the U.S. Supreme Court. Of course, many other states and municipalities are interested in the outcome and are siding with Albuquerque. This is especially true of other western states with Indian reservations that might attempt to enforce their own environmental standards, as Isleta Pueblo is doing.

Kevin Mitnick and Computer Hacking

On February 15, 1995, perhaps the most unusual manhunt in history ended when federal agents arrested Kevin Mitnick in Raleigh, North Carolina. Mitnick was accused of breaking into computer systems around the world, sometimes altering information, but always causing the owner of the computer great expense and trouble in setting up new security measures to keep him out. Mitnick was also charged with violation of the conditions of his probation from previous convictions for computer crime.

Mitnick had been on the run for two years. The FBI had been trying unsuccessfully to locate him by tracing his access to the Internet. Mitnick wasn't finally caught until he made the mistake of breaking into the San Diego Supercomputer Center. His hacking led him to some of the files of computer security expert Tsutomu Shimomura. This hacking angered Shimomoru who agreed to help the FBI track Mitnick down. For two weeks, Shimomoru helped search for Mitnick, finally tracing him to an apartment complex in Raleigh. Mitnick's communications were monitored, and at 1:30 AM, when he went on-line, federal agents moved in and arrested him.

Interestingly, in all of his computer hacking, Mitnick never seems to have done anything for personal financial gain. He seems to have lived quite modestly. He did, however, defraud telephone companies of many hours of long-distance time. Mostly, he seems to have hacked for the challenge and sometimes to exact revenge on people.

Kevin Mitnick has a long history of computer abuse. Like many teenagers, he became fascinated with computer technology while taking a computer class in high school in the Los Angeles area. He learned quickly and soon managed to hack into the L.A. public school district's computer system. At age 17, Mitnick hacked into the computer system of Pacific Bell and altered phone bills. That year, he also accessed information valued at $200,000 from a San Francisco company. He was prosecuted for these crimes through the juvenile system and received six months probation. While on probation, he hacked into the telephone company again and had his probation officer's phone disconnected, and he accessed a credit service computer and altered the computerized credit record of the judge on his case. Police found that his police records had been accessed from the outside. During this time, he was also convicted of stealing software from a company in Santa Cruz, California. Interestingly, although this conviction showed up in a federal criminal database, there was no record of this conviction in the police computer files in Santa Cruz.

Mitnick also hacked into the computers of the North American Air Defense Command (NORAD) in Colorado Springs. NORAD is part of an early warning system and is responsible for monitoring possible missile attacks directed towards the United States and coordinating the response of the U.S. military. Obviously, damage to these files or the implantation of false information could have catastrophic effects.

In 1988, Mitnick was charged with two new crimes: causing $4 million in damage to a Digital Equipment Corporation computer while stealing a secret computer security system and accessing the MCI network to make free long-distance phone calls. This time, the judge decided that Mitnick was dangerous to

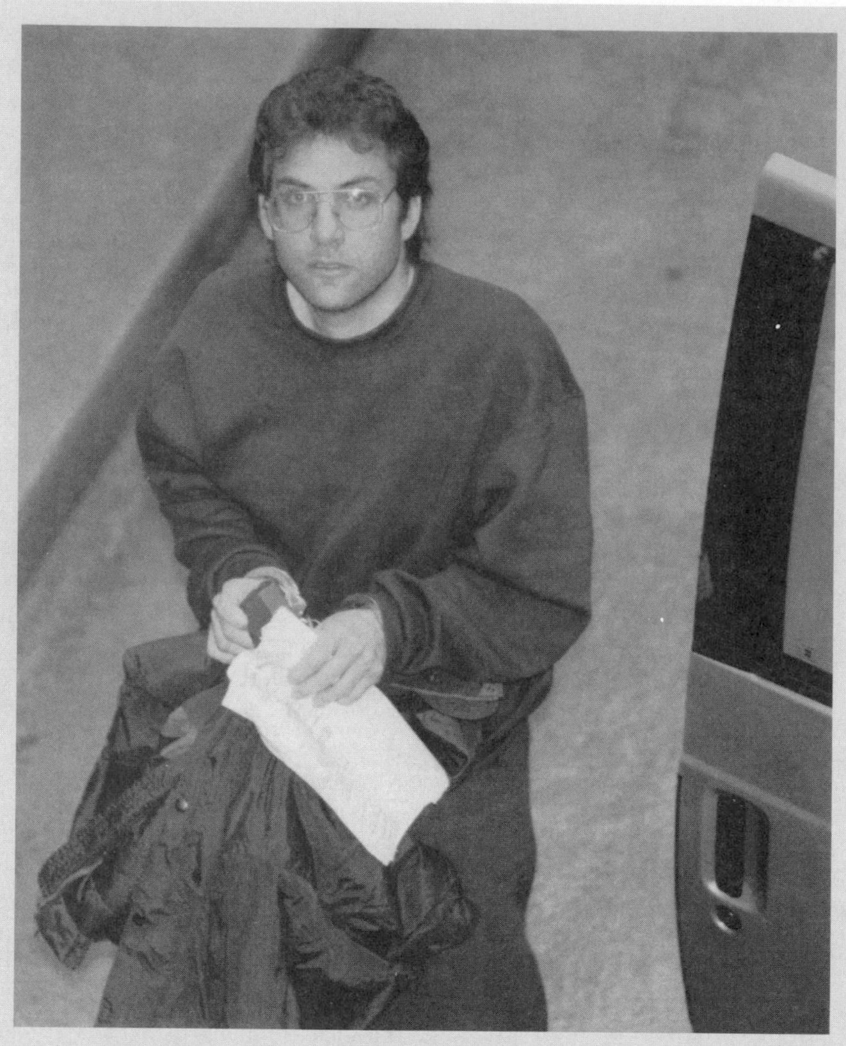

Kevin Mitnick, whose wide ranging computer hacking has led to his arrest on several occasions for computer crime. Photo by Bob Jordan - AP/Wide World Photos.

society and ordered him held in jail without bail. He was charged under a new federal law that made it a crime to gain access to an interstate computer network for criminal purposes. During his attempts to be released on bail, prosecutors presented evidence that he had penetrated a National Security Agency computer in Washington. He had also apparently planted a false story on a financial newswire, reporting that Security Pacific Bank had lost $400 million in the first quarter of 1998. This story was planted just four days after Mitnick had been turned down for a job at Security Pacific. Fortunately, the false story was detected and killed before it caused harm to the company and its investors. On March 15, 1989, Mitnick pleaded guilty to the two counts and was ultimately sentenced to one year in prison, six months in a residential psychological counseling program, and three years of probation.

In 1992, the FBI suspected Mitnick of hacking into the computers at Pacific Bell again and searched the business in Las Vegas where he was then working. After this search, Mitnick disappeared. While investigating the Pacific Bell break-in, the FBI found evidence that Mitnick had hacked into the California driver's license records to try to set up a false identity for himself and had even posed as a police officer to try to get driver's license information. A warrant for his arrest was issued, and the manhunt was on. Police came close to arresting him several times, including once when he went to a copy shop to pick up a Fax of information illegally obtained from the driver's license department. At one point, he was traced to Seattle, but he moved on before being arrested.

After Mitnick was finally found in Raleigh and arrested in 1995, he was arraigned for a host of new computer-related crimes and for violation of his parole. He was denied bail on the grounds that it would be impossible to keep him from committing more computer crimes while waiting trial if he were free in the community. The authorities had to go to great lengths to ensure that he couldn't continue to commit crimes from jail. He was denied access to any computer equipment, and all of his phone calls were closely monitored to prevent him from accessing a computer remotely. In 1997, Mitnick was sentenced to 22 months in prison on charges of hacking into the phone system to get cellular phone codes and numbers that he used for free long-distance access to computer networks. This sentence is to be followed by three years of probation during which he must stay away from anything "high tech," including computers and telephones, unless he has the consent of his probation officer. However, this probation may never occur, since he still faces numerous other charges related to computer crime. There are probably many break-ins perpetrated by Mitnick that have yet to be discovered.

This case illustrates the incredible potential for misuse of computers that was described previously in this chapter. What is the engineer's responsibility for the ethical use of computers? Clearly, as people who are well trained in high technology, the engineer can have a unique access to and knowledge of computer systems. Yet, our professional ethics demands that we put this knowledge to good use. It is also important that those who design computers and their operating systems keep this type of activity in mind and attempt to design so that opportunities for misuse can be minimized.

PROFESSIONAL SUCCESS: FALSIFYING EXPERIMENTAL RESULTS

Experimental work is an important part of an engineering student's education. It is no surprise that ethical issues often arise in the course of laboratory work. Most ethical issues in experimentation relate to honesty in reporting results. For example, it is often tempting to "massage" data to get the desired result. Or sometimes, it seems easier to "dry run" an experiment by recording measurements and results in your lab book even though you haven't actually performed the experiment. Fundamentally, these are very similar to cheating.

How do you decide what is ethical in experimentation? It is easiest to look at ethical issues related to experimentation using virtue ethics just as we did in the box on cheating. There we saw that honesty is a virtue that should be fostered within ourselves. So, virtue ethics tells us that the utmost care must be taken to ensure that experiments are performed carefully and the results reported honestly.

KEY TERMS

Whistleblowing
Computer ethics
Conflict of interest
Environmental ethics

REFERENCES

Charles E. Harris, Jr., Michael S. Pritchard, and Michael J. Rabins, *Engineering Ethics, Concepts and Cases*, Wadsworth Publishing Company, Belmont, CA, 1995.

Mike W. Martin and Roland Schinzinger, *Ethics in Engineering*, 2d. ed., McGraw-Hill, New York, 1989.

Boris Rauschenbakh, "Computer War," in *Breakthrough; Emerging New Thinking*, A. Gomyko and M. Hellman, Eds., Walker, New York, 1988.

BART

Robert M. Anderson, *Divided Loyalties*, Purdue University Press, West Lafayette, IN, 1980.

Gordon D. Friedlander, "The Grand Scheme," *IEEE Spectrum*, Sept. 1972, p. 35.

GORDON D. FRIEDLANDER, "BART's Hardware—from Bolts to Computers," *IEEE Spectrum*, Oct. 1972, p. 60.

GORDON D. FRIEDLANDER, "More BART Hardware," *IEEE Spectrum*, Nov. 1972, p. 41.

GORDON D. FRIEDLANDER, "Bigger Bugs in BART?," *IEEE Spectrum*, March 1973, pp. 32–37.

GORDON D. FRIEDLANDER, "A Prescription for BART," *IEEE Spectrum*, April 1973, pp. 40–44.

Goodrich A7-D Brake

JOHN FIEDLER, "Give Goodrich a Break," *Business and Professional Ethics Journal*, vol.7, Spring 1988, p. 21.

KERMIT VANDIVIER, "Why should my conscience bother me?", in Robert Heilbroner, Ed., *In the Name of Profit*, Doubleday, 1972.

KERMIT VANDIVIER, "The Aircraft Brake Scandal," *Harpers Magazine*, April 1972, p. 45.

TANIA SOUSSAN, "8 States Watching Water Quality Suit" *Albuquerque Journal*, Section C, June 24, 1997, p. 1.

Isleta Pueblo Water

TANIA SOUSSAN, "Isleta's Water Demands Upheld" *Albuquerque Journal*, Section A, Nov. 11, 1997, p. 7.

Computer Hacking

KIM MURPHY, "Ex-Computer Whiz Kid Held on New Fraud Counts," *Los Angeles Times*, December 16, 1988, Part II, p. 1.

JOHN JOHNSON, "'Dark Side' Hacker Seen as 'Electronic Terrorist'" *Los Angeles Times*, January 8, 1989, Part II, p. 1.

JOHN JOHNSON AND JULIE TAMAKI, "The Fugitive Hacker," *Los Angeles Times*, July 12, 1994, Part B, p. 1.

JOHN JOHNSON, "A Cyberspace Dragnet Snared Fugitive Hacker," *Los Angeles Times*, February 19, 1995, Section A, p. 1.

Problems

1. An engineer leaves a company and goes to work for a competitor.
   a. Is it ethical for the engineer to try to lure customers away from the previous employer?
   b. Is it alright for the engineer to use proprietary knowledge gained while working for the previous employer at the new job? How would the answer to this question change if the new job weren't for a competitor?
   c. At a new job, is it acceptable to use skills developed during your previous employment?
2. If you are an engineer working for a state highway department with the responsibility for overseeing and regulating construction companies that work for the state, is it a conflict of interest to leave the state and accept a position with a construction company that you formerly regulated as a government relations manager? Is the opposite acceptable: leaving a private company to take a position in government regulating that company? How about if you have substantial stock in the company in a pension or other plan?
3. You are an engineer who has taken a new job with a competitor of your previous company. At a meeting you attend, a research engineer describes her plans for developing a new product similar to one developed by your former company. You know that the direction this engineer is taking will be a dead end and will cost the company a lot of time and money. Do you tell her what you know? Does the answer to this question change if the new company is not a direct competitor of the previous one?
4. You are a civil engineer working for an engineering consulting firm and have just finished work on a new bridge project. This project involved some innovative designs developed by you and other engineers in the firm. You have decided that you now have enough experience to start your own consulting firm. The first project that comes to you is a bridge. Can you use

the innovation pioneered at your previous firm in this new design? How does this situation differ from that in Question 3?

5. Write a code of ethics for computer use.
6. Is there an ethical obligation to ensure that the information you post on your Internet Web site is accurate and true? Or is it up to the Web user to be discriminating and to realize that some material might not be accurate?
7. There is much in the news about the use of the Internet to disseminate pornographic images, especially in the context of the availability of this material to children. What ethical issues does "cyberporn" and efforts to limit it raise? Do employers have the right to fire employees who access pornography on their computers at work?
8. Many desktop computers come with games already installed on them. In addition, there are many Web sites where users can download games onto their computers. Is it alright to play computer games at work? How about during lunch? After hours?
9. Should computer and software designers be concerned about possible abuse of their products? Should designs incorporate methods for preventing the misuse of computers?
10. Is it acceptable for employees to use their computers at work to send and receive personal e-mail?
11. There has been some discussion of having the federal government maintain a computer database of medical information on everyone in the United States. Some medical researchers feel that such a database might save lives by allowing access to a larger base of medical records for research purposes. Certainly, this database would make certain legitimate government functions more efficient. Is this a good idea?
12. Does Roger Boisjoly's testimony in front of the presidential commission described in Chapter 1 constitute whistleblowing?

BART

13. BART was a very innovative design that went well beyond other mass transit systems then in existence. What guidance does "accepted engineering practice" provide in such an innovative design?
14. When pointing out safety problems, an engineer is rightfully concerned about maintaining his job. However, how effective is an anonymous memo? Can anyone be expected to pay attention to something that a person won't sign?
15. Did the three engineers meet the criteria for whistleblowing discussed previously in this chapter?
16. Should the IEEE have intervened in the court case?
17. In what ways could the BART structure and chain of command have been changed to make the whistleblowing unnecessary?
18. At what point should an engineer give up expressing her concerns? In this case, when several levels of management appeared not to share the engineers' concerns, how much more effort does professional ethics dictate is necessary?
19. What level of supervision should an organization have over its contractors? Is it sufficient to assume that they are professional and will do a good job?
20. One of the perceived problems with BART was a lack of adequate documentation from Westinghouse. What are the ethical considerations regarding the documentation of work? What responsibility does an engineering organization have after the design is complete?
21. It is important to remember that from our perspective, it is impossible to know whether the Westinghouse test procedures and schedule were adequate. The subsequent accidents and problems really don't tell us much about this issue: Anything new and this complex should be expected to have some bugs during the early periods of operation. Given this understanding, were the engineers' concerns adequately addressed by management? What actions short of going to the board and whistleblowing might the engineers have taken?

Goodrich A7-D Brake

22. Was an unethical act taking place when test results on the brake were falsified?

23. Was this mitigated at all by the fact that Goodrich was planning to redesign the brakes anyway?
24. Was this mitigated by the fact that the brake design was a new one for which the old test methods might not be applicable? This was a claim by Goodrich. If the old test methods were not applicable to the new design, what should Goodrich have done?
25. Can some of the problem here be attributable to sloppy management? For example, should the original designer be allowed to hand off the test work to a new hire with no further participation? What are the ethical implications of this type of management?
26. Did Vandivier meet the criteria set out in the previous section for whistleblowing? In other words, was there need for the whistle to be blown? Did he have proximity? Was he capable? Was it a last resort? Does the fact that nothing seems to have been done to Goodrich following the Senate investigation change your answer?
27. What could Goodrich have done to solve the problem without public disclosure of the falsified tests?
28. Was Goodrich engaged in a "bait and switch?" In other words, did it use claims about the innovative brake design as a means to get the contract, with the intent of ultimately supplying a conventional brake? What is the ethical status of this type of tactic?

Isleta Pueblo Water

29. What does utilitarianism tell us about this case? What do rights and duty ethics say?
30. Is the religious use of water a valid claim against a municipality? If this were a claim by a large mainline religious denomination (for example, some Christian denominations might want to use the river for baptisms), does the answer to this question change the answer?
31. It might be argued that the pueblo benefits economically by its proximity to the city. Does this aspect affect your answers to Question 1?
32. Engineers frequently participate in setting standards for pollution limits through consulting with governments. If you were an engineer working for the EPA, what would your advice be? How would this advice change if you worked for the city of Albuquerque? How would this advice change if you worked for Isleta Pueblo?
33. Nearly every municipality in the United States has some pollution problems and controversies. Research a local pollution issue and apply the problem-solving techniques discussed in this book to determine what you think is the ethical solution to the problem.

Computer Hacking

34. Is hacking into a computer system just to look around ethically acceptable?
35. In his defense, Mitnick has claimed that he never enriched himself through his hacking. Can this claim be believed? If true, is it significant in his defense?
36. What ethical responsibilities do engineers and software designers have for ensuring that hacking is minimized?

7

Ethics in Research and Experimentation

In the spring of 1989, the scientific community was electrified by the announcement that two researchers in Utah had discovered a way to harness nuclear fusion in a small electrochemical cell. If true, this discovery would revolutionize the production of energy, perhaps leading to small power-generating plants in every home.

Conventional approaches to initiating fusion reactions involved heating gases to extremely high temperatures or bombarding a specially prepared target with energetic beams to start a nuclear reaction. These methods involved very large and costly equipment and had yet to produce more energy than was required to initiate the reaction. In contrast, the new process, which was termed "cold fusion," was simple, inexpensive, and seemed to produce copious amounts of energy. If verified, this new innovation would also turn the conventional wisdom about how nuclear reactions proceed inside out. In the aftermath of this announcement, scientists and engineers at research labs and universities all over the world tried to duplicate the experiment.

One group trying to reproduce these results was at Texas A&M University. During the course of their work, they started to see some cells that contained tritium, an expected byproduct of fusion that had been absent in the

Sections

Objectives

After reading this chapter, you will be able to:

- Determine what ethical issues arise in the course of research and experimentation.
- Decide which analysis methods are most applicable to ethical issues in research.
- Learn about pathological science and ways to avoid it.

Utah experiments. The detection of tritium would lend credence to the theory that fusion was indeed taking place in the cells. Curiously, only a few of the Texas A&M samples contained tritium. The lack of tritium in the Utah experiments and the sporadic detection of tritium in the Texas A&M work should have caused the researchers to be careful about reporting their results. Instead, caught up in the excitement of the discovery, they quickly reported their results at conferences and in the scientific literature. Soon, rumors and accusations of shoddy work or even deception began to circulate within the scientific community. This ultimately hurt the reputation of the researchers—both students and faculty—and the university.

During the course of their careers, engineers are frequently involved in research, experimentation, or the testing of new products. These activities bring with them some new ethical issues that we haven't yet considered. In this chapter, we will look at the special ethical challenges presented by research and see how these situations can be navigated ethically.

7.1 INTRODUCTION

Many engineers will become involved in research and experimentation in the course of their academic and professional careers. Even engineers who are not employed in research laboratories or academic settings can be involved in research and development work or the testing of a new product or design. Although the ethical ideas discussed previously in this book are also applicable here, experimentation and research present some ethical challenges that warrant special consideration.

7.2 ETHICS AND RESEARCH

There are two major ethical issues related to research: honesty in approaching the research problem and honesty in reporting the results. The first relates to a state of mind essential to successfully performing research. This state of mind includes avoiding preconceived notions about what the results will be, being open to changing the hypothesis when such action is warranted by the evidence, and generally ensuring that an objective frame of mind is maintained. As we will see in the cases at the end of this chapter, this attitude is not necessarily easy to assume, but it is essential to producing useful research or test results. More will be said about this topic later in this chapter in the section on pathological science.

Results must also be accurately reported. Once an experiment or test has been performed, the results of the experiment must not be overstated, but rather an accurate assessment and interpretation of the data must be given. The environment that most researchers work in fosters temptations and rewards for overstating research results. Academic researchers must publish significant research results in order to get tenure at their universities. If an experiment isn't working out, it is tempting to "massage" the results to achieve the desired outcome. Even for researchers in industrial environments or faculty who are already tenured, the quest for fame or the desire to be the first with new results can be overwhelming and can lead to falsification of data. Often, the pressure to get a new product to market leads the test engineer to "fudge" data to qualify the product.

It is important to note the distinction between intentional deception and results or interpretations that are simply incorrect. Sometimes, results are published that, upon further research, turn out to be incorrect. This situation is not an ethical issue unless a clarification of the results is never presented. Rather, this issue indicates that great care must be taken before results are initially reported.

It is also important to ensure that proper credit is given to everyone who participated in a research project. Rarely is research performed by a single investigator working alone in her laboratory. Generally, there is participation by other people, who should be acknowledged for their contributions such as discussions or guidance, construction of experimental apparatus, or substantial help with performing experiments or interpreting data.

It is tempting to think that fraud and deception in research are rare and only perpetrated by lower level scientists, but this perception is decidedly untrue. There are many examples of well-known and even Nobel prize–winning scientists who have had lapses of ethical judgement with respect to their research. For example, Robert Millikan was a physicist from the University of Chicago who won the 1923 Nobel Prize in physics for experiments that measured the electrical charge of the electron. Studies of Millikan's unpublished data indicate that he excluded 49 of the 140 experimental observations from the paper that he published [Holton, 1978, and Franklin, 1981]. However, in the paper, he stated that the published work contained all of the data. Inclusion of this data wouldn't have changed his conclusions, but would have made the result seem more certain and the experiment more clearly definitive.

Analyzing Ethical Problems in Research

How can ethical issues relating to research best be analyzed? Perhaps the easiest means to determine the best ethical course in performing research and experiment is to consult the codes of ethics of the engineering professional societies. All of the codes include language requiring engineers to be honest in reporting the results of work and assigning credit for work done. For example, the code of the American Institute of Chemical Engineers states that "members shall... treat fairly all colleagues and co-workers, recognize the contributions of others . . . ," and "issue statements and present information only in an objective and truthful manner." These statements apply equally well to all professional activities of an engineer, including research, experiment, and testing.

The ethical theories discussed in Chapter 3 can also be used to analyze issues involving research. Utilitarianism or rights and duty ethics can be applied to research, but it is perhaps easiest to examine research issues using virtue ethics. As discussed in Chapter 3, one of the virtues is honesty. Honesty facilitates trust and good relations between individuals, whereas dishonesty leads to doubts and misgivings about others. People rarely want to associate with those who they feel don't behave fairly and can't be trusted. Making false claims about the results of experiments is certainly a form of dishonesty. We should seek to enhance virtues such as honesty within ourselves and others, so virtue ethics clearly tells us that the inaccurate reporting of experimental results is unethical. Likewise, not giving credit to everyone who has participated in a project is dishonest, and virtue ethics indicates that this practice is unacceptable.

7.3 PATHOLOGICAL SCIENCE

As mentioned previously, self-deception is one of the biggest impediments to the successful completion of a research or experimental project. Self-deception in research is a frequent occurrence in many areas of science and has led to some notorious cases throughout history. Irving Langmuir, a well-known physicist working at General Electric Research Laboratories, coined a term for this phenomenon: "pathological science." He proposed the following six characteristics of pathological science [Langmuir, 1968]:

1. The maximum effect that is observed is produced by a causative agent of barely detectable intensity, and the magnitude of the effect is substantially independent of the intensity of the cause.

This characteristic implies that it doesn't matter how close the causative agent is or how intense it is; the effect is the same. This practice, of course, goes against all known forces and effects.

2. The effect is of a magnitude that remains close to the limit of detectability; or, many measurements are necessary because of the very low statistical significance of the results.

   The problem here is that when things are at the edge of statistical significance or of detectability, the tendency is to discard values that don't "seem" right. To measure anything at the edge of detectability requires a lot of data. With a lot of data to work with, the measurements can be massaged to fit the conclusion that is being sought. In fact, what often happens is that data is rejected based on its matching of the preconceived theory, rather than on its true significance.
3. Claims of great accuracy.
4. Fantastic theories contrary to experience.
5. Criticisms are met by ad hoc excuses thought up on the spur of the moment.
6. Ratio of supporters to critics rises up to somewhere near 50% and then falls gradually to oblivion.

The term "pathological science" doesn't imply any intentional dishonesty, but only that the researcher comes to false conclusions based on a lack of understanding about how easy it is to trick yourself through wishful thinking and subjectivity.

This shows that a great deal of objectivity and care in the pursuit of research or testing is required. Drawing conclusions on very subtle effects is very tricky, and these conclusions should be confirmed by as many colleagues as possible. Ultimately, the goal of research is not publicity and fame, but rather the discovery of new knowledge.

APPLICATION: CASES

The N-Ray Case

After the discovery of X-rays in the late 19th century, there was a great deal of interest among scientists in finding other similar types of rays. Many scientists joined this search in the hopes of achieving the fame that such a discovery would bring. In many ways, this scenario was similar to the frenzy in the scientific community in the 1980s upon the discovery of superconductivity at temperatures above the boiling point of liquid nitrogen. Many researchers dropped everything else they were doing and began searching for new materials with even higher superconducting temperatures, especially hoping to find one at room temperature. The search to find new rays was joined by a well-known French physicist, René Blondlot, at the University of Nancy. His case is discussed in depth in an interesting article published in 1980 by *Scientific American* [Klotz, 1980].

The apparatus used at the time for detecting such rays was the spark gap. This device consisted of two electrodes that were close enough together so that a spark developed between them in air when a large electric potential was applied between them. What we now know as electromagnetic radiation in the form of light or X-rays directed through the spark gap increased the ionization in the gap, increasing the current flow and the brightness of the spark. The brightness of the spark couldbe used to measure the intensity of the radiation present in the gap. Of course, by modern standards this is a very crude means for detecting X-rays, but at the time, this method was state of the art.

In order to see the change in brightness, care had to be taken in establishing the measuring environment. For example, the researcher had to stay in a darkened room sufficiently long so that his eyes would become dark-adapted. Even then, the change in intensity of the spark could be very subtle, and care had to be taken to be honest in the assessment of the change.

In 1903, Professor Blondlot was working with gas discharges that produced the newly discovered X-rays. His previous experience was in the study of electromagnetic phenomena, and he was hoping to discover if X-rays were a wave or particle by determining if the X-rays could be polarized as visible light can be. Using a spark gap and an apparatus similar to the one that Roentgen had used to discover X-rays, Blondlot attempted to determine the polarization of X-rays by rotating the spark gap in the X-ray field. In his initial study, Blondlot discovered that, indeed, the spark gap became brighter when rotated to a certain angle with respect to the discharge tube. This was an important discovery.

Subsequent experiments indicated that the radiation impinging on the spark gap could be bent by a quartz prism. This feature was a major problem, since X-rays had already been shown by many scientists to be unaffected by lenses and prisms. The fact that the radiation he was measuring appeared to be bent by the prism convinced Blondlot that he had discovered a new form of radiation that he called N-rays (for the University of Nancy). He quickly published this work.

The reports of the discovery of a new type of ray set off a flurry of activity in other laboratories around the world, and Blondlot himself continued to study the phenomenon. Many discoveries were made about N-rays: Materials were found that transmitted them (metals, wood, mica, quartz) and some that didn't transmit the rays (water and rock salt). Natural sources of N-rays were also discovered, including the sun and the human body. Despite the explosion of research on N-rays, there were also some doubts about Blondlot's findings. Many researchers outside of France, including Lord Kelvin in England, had been unable to reproduce the results reported by Blondlot.

Prof. J. W. Wood of Johns Hopkins University was also unable to reproduce the results and traveled to Nancy to observe the experiments firsthand. In a paper published in *Nature*, he described the experiments that he had witnessed. Wood reported that when he observed the spark gap and someone placed a hand in the path of the N-rays, Wood didn't see the expected changes in intensity. Told that his eyes weren't sensitive enough, he exchanged positions with the French researchers and placed his hand in the path. The research team incorrectly reported whether his hand was in or out of the beam as they claimed to see changes in intensity. Wood reported that there was no correlation between the position of his hand and their reports of intensity.

Wood also observed a different experiment designed to spread the N-rays into a spectrum. The dispersion of the N-rays was accomplished using an aluminum prism and was observed using a thin phosphor strip painted onto a cardboard screen. Wood was unable to observe the variations in intensity from the phosphor that the French team claimed to see. Indeed, when Wood surreptitiously removed the prism from the apparatus, the researchers still claimed to see the effect! Wood was convinced after this incident that there were no N-rays and that the researchers had deluded themselves.

Publication of Wood's findings ended research into N-rays everywhere except in France. Blondlot responded to the criticisms and continued to present results of new, more controlled experiments. He even published a set of instructions for properly observing the phenomenon. For example, the instructions stated that the observer had to avoid gazing directly at the spark gap and instead had to look at it obliquely. The observer had to remain silent, avoid smoke, and had to look at the detector in the "way an impressionist painter would view a landscape" [Klotz, 1980]. Acquisition of this ability required a great deal of practice and might be impossible for some people. In other words, the key to the measurement was the sensitivity of the observer, rather than the validity of the phenomena. As more research was performed, it became clear even to the French that there were no N-rays.

The Case of Cold Fusion at Texas A&M University

On March 23, 1989, Stanley Pons and Martin Fleischmann of the University of Utah announced that they had produced excess heat in a tabletop electrochemical cell. The excess heat was presumed to be due to nuclear fusion, and the process was dubbed "cold fusion." Pons'

Stanley Pons and Martin Fleischmann, who started a frenzy in the scientific research community when they announced that they had discovered a way to control nuclear fusion in a table-top electrochemical cell. AP/Wide World Photos.

and Fleischmann's results were widely reported in newscasts and daily newspapers and led to great excitement among scientists around the world.

The apparatus used by Pons and Fleischmann was a fairly standard electrochemical cell. They found that when palladium electrodes were immersed in heavy water (water with the normal hydrogen atoms replaced by the heavier deuterium isotope) and an electric current run through them, heat far in excess of levels expected was produced. This heat production was attributed to the breakdown of the heavy water due to electrolysis, diffusion of the deuterium into the palladium, where the deuterium was thought to get to a density sufficient to initiate fusion, leading to the release of the excess heat.

Although Pons and Fleischmann were well-respected electrochemists, their results were treated with great skepticism by many scientists, especially those who had worked in conventional fusion and nuclear physics. This skepticism arose because, according to the contemporary understanding of the fusion process, the reaction of deuterium should produce copious amounts of tritium (another hydrogen isotope) and neutrons. Neither of these products was seen in the Pons–Fleischmann experiments. The response of many of the believers in cold fusion to this criticism was that they had discovered some new form of fusion that didn't behave according to the old rules. Indeed, there were some claims of professional jealousy: Physicists who had worked for years to make conventional fusion practical would not be happy to be upstaged by chemists who couldn't possibly know anything about fusion. Despite the controversy, the potential benefits if this process proved to be real were so enormous that many researchers worldwide began setting up similar electrochemical cells in their laboratories and trying to reproduce the results. John Bockris at Texas A&M University was one of these scientists.

Bockris' research group built electrochemical cells like those of Pons and Fleischmann and set out to verify the Utah work. By April 22, 1989, this group had observed a surprising result. A graduate student working with Bockris, Nigel Packham, had removed samples of the electrolyte from three of the cells in the laboratory and took them to another campus building, the Cyclotron Center, for tritium measurements. Two of the three samples were "hot," containing 10^9 tritium atoms/ml, an amount far in excess of the expected background level. Subsequently, tritium was detected in four more cells.

When this work was reported at scientific meetings, there was immediate concern, since the data was too amazing. More work was performed, designed to control the experimental conditions more carefully, including work by other researchers at Texas A&M. For example, Kevin Wolf, a nuclear chemist, ran a cell in front of neutron detectors in his laboratory, hoping to find the telltale sign that should accompany tritium production. No neutrons were detected, although tritium did appear in the electrolyte when tested. Packham also performed an experiment in which electrolyte samples were taken at four different intervals over twelve hours while the cell was running. At the beginning of the experiment, tritium was at background levels. Two hours later, it was slightly above background level. A few hours later, the level had climbed greatly to 5 trillion atoms, and at 12 hours it had climbed to 7.6 trillion atoms [Taubes, 1990]. Although this data seemed to confirm that tritium was being produced in the cell, skeptics also pointed out that this result was consistent with someone "spiking" the sample with tritium sometime toward the middle of the experiment. Indeed, there was a supply of tritium stored in the laboratory.

In response to these allegations, Bockris and his team failed to take steps to ensure that intentional spiking couldn't occur. Offers to place the experiment in the locked laboratory of colleague Charles Martin, another electrochemist, were refused, and the bottle of tritiated water in the lab was not locked up or thrown away. While Bockris continued his work, Wolf and Martin continued their own similar studies with the same type of cell used by Bockris. Martin even took the precaution of taking cells home to ensure that there would be no sabotage. Martin's cells never showed signs of tritium.

In late September, after nearly three months with no results, two more cells turned up with tritium. The discovery of new cells containing tritium coincided with a scheduled visit from officials of the Electric Power Research Institute (EPRI), which had funded some of the research at Texas A&M. This incident and coincidences with other visits from funding sources cast more suspicion on the tritium results.

On November 27, 1989, Packham, who had not been involved with this work for several months, decided to test samples from two previously untested cells with titanium electrodes. These samples proved to be hot as well. The coincidence was too much for several of the workers in the lab. They took their concerns to Bockris, who dismissed their claims. These scientists subsequently went to other laboratories or sought employment outside the university.

Through most of the controversy, the university had taken a hands-off approach. There had been inquiries of Bockris as to his results and why they appeared so anomalous. However, the university allowed the situation to continue. In June of 1990, Gary Taubes published an article on this situation in *Science*. The negative publicity, especially the statements that the university appeared to be doing nothing, prompted an internal investigation by the university. The three-member panel appointed by the university concluded that intentional spiking of the samples could not be ruled out, but that it was more probable that the results were due to inadvertent contamination or other unexplained problems with the measurements. The panel did find that there were lapses in proper scientific procedure caused by the excitement surrounding the study of a new discovery that was receiving so much media attention

tion. These lapses included categorizing experiments that supported the hypothesis of cold fusion as "successful" and those that didn't support it as "failures" [Pool, 1990].

Unable to reproduce the Pons–Fleischmann results, many researchers stopped their investigations of cold fusion. Funding for this work has dried up, although there are still a few people who believe in the phenomenon and continue to study it. Fraud was certainly a possibility at Texas A&M, although it is unclear who was responsible if this is true. However, all of the researchers were responsible for performing their experiments in an objective manner. In the face of charges of fraud, steps should have been taken to ensure that spiking was not possible. The reputations of senior scientists as well as of students and the university were tarnished by this episode.

KEY TERMS

Ethics in research
Pathological science

REFERENCES

WAYNE LEIBEL, "When Scientists are Wrong: Admitting Inadvertent Error in Research," *Journal of Business Ethics*, vol. 10, 1991, pp. 601–604.

ALEXANDER KOHN, *False Prophets*, Basil Blackwell, Oxford, 1986.

GERALD HOLTON, "Subelectrons, presuppositions, and the Millikan-Ehrenhaft dispute," *Historical Studies in Physical Sciences*, vol. 11, 1978, p. 185.

A. D. FRANKLIN, "Millikan's published and unpublished data on oil drops," *Historical Studies in Physical Sciences*, vol. 58, 1981, p. 293.

IRVING LANGMUIR, "Pathological Science," in R. N. Hall, Ed., General Electric Research and Development Center Report No. 68-C-035, April 1968.

N-rays

IRVING M. KLOTZ, "The N-ray Affair," *Scientific American*, May 1980, vol. 242, no. 14, pp. 168–70.

Articles in *New York Times*, March 24 and 28, 1989, and numerous subsequent articles. Also, this story was widely reported at the same time in many daily newspapers across the country.

Cold Fusion

ROBERT POOL, "Cold Fusion at Texas A&M: Problems, But No Fraud," *Science*, 14 December 1990, vol. 250, pp. 1507–8.

GARY TAUBES, "Cold Fusion Conundrum at Texas A&M," *Science*, 15 June 1990, vol. 248, pp. 1299–1304.

Problems

1. How can utilitarianism or rights and duty ethics be applied to issues surrounding the proper conduct of research?
2. Think about ways in which ethical issues regarding experimentation or research have come up during your academic career. Analyze these cases and decide if you handled them ethically or not.
3. Read the papers by Holton and Franklin listed in the references and obtain a copy of Millikan's original paper. Do you think that there are ethical problems in Millikan's actions?
4. As described in the case on the A7-D brakes at the end of Chapter 6, test results on brakes were falsified in order to meet a tight schedule for qualification. What do the engineering codes of ethics say about this situation? How should this situation have been handled ethically?

N-rays

5. How well is the N-ray case described by Langmuir's six characteristics of pathological science?
6. What ethical mistakes were made by Blondlot and his colleagues in researching N-rays?
7. Once Wood's article was published in *Nature*, what should Blondlot have done?

Cold Fusion

8. How seriously should Bockris have taken the suggestions or charges of fraud made against his results? How seriously should the university have taken these charges?
9. Do some further reading on the early claims about cold fusion. Analyze the claims of Pons and Fleischmann in terms of the description of characteristics of pathological science.
10. Using the code of ethics of one of the professional societies printed in Appendix A, analyze the behavior of Bockris' research group. According to the code, did this group operate in an ethical manner?

8

Doing the Right Thing

Most of the cases that we have studied thus far, and indeed most of the cases presented in engineering ethics studies, are retrospective looks at disasters or wrongdoing. These cases all have in common that something bad happened: A mistake was made in a design, pressures were put on engineers to make bad decisions, or illegal and immoral activity was being covered up. The purpose of these cases is to see what went wrong and to understand how to go about doing the right thing when faced with similar circumstances.

In contrast, the two cases that will end this book do not involve disasters, but rather are examples of how things should be done in the first place to avoid disasters. Here we will see that sometimes when a design flaw is noticed, even after the design has long since been implemented, everyone can cooperate to do the right thing rather than point fingers at each other, deny responsibility, and immediately head to court.

Objectives

After reading this chapter, you will be able to:

- See how ethical problems can be avoided.
- Learn how engineers can cooperate with each other and with clients and government agencies to be sure that the ethically correct choice is made.

The Citicorp Center in Manhattan (white building in center top). Structural problems in the design of this building led to a retrofit to ensure that the building could withstand extremely high winds. Photo by Marc Anderson - Simon & Schuster/PH College.

APPLICATION: CASES

The Citicorp Center Case

In the early 1970s, planning started for a new headquarters for Citicorp in Manhattan. The new building, to be called Citicorp Center, would take up an entire city block. But the site chosen was problematic, since one corner of the block was occupied by a church that had been built in 1905. In order to acquire the site, Citicorp agreed to demolish the old church and build a new freestanding church as part of the Citicorp Center. To accomplish this task, the Center's architect Hugh Stubbins, Jr. and structural engineer William LeMessurier designed a 59-story tower that was set on four large nine-story-high columns. These columns were placed in the middle of each side of the building rather than at the four corners. This arrangement allowed the church to be built beneath the tower, one corner of which was cantilevered over the church.

The design of a skyscraper such as the Citicorp Center involves many different professionals. Perhaps most important is the structural engineer, whose job it is to ensure that the building's superstructure will be strong enough to hold the building up and will withstand the forces of nature, especially wind. LeMessurier designed a unique system of wind braces for the building calling for 48 chevron-shaped steel members welded together to form the superstructure.

Four years after the Citicorp Center had been built, a question from an engineering student led LeMessurier to look at his design again. His new calculations showed that under some wind conditions, the forces that the braces had to withstand were about 40% larger than his original calculations had shown. Normally, this wouldn't have been a problem, since even with the extra stresses, the building would have been strong enough to withstand the expected loads. However, just weeks before, he had learned that during construction, the welded joints in the superstructure had been replaced with bolted joints. This replacement was done with the approval of engineers from his firm. In light of his new calculations, LeMessurier was concerned about whether the bolted joints would have the necessary strength to withstand strong winds. Further calculations and testing showed that his concerns were well founded and the bolted joints would be dangerously weak when the building was subjected to strong winds. How strong a storm would be required to cause the building's structure to fail? Meteorological records for New York indicated that a storm with sufficiently strong winds to tear the joints apart could be expected on average once every 16 years. LeMessurier quickly developed a plan to solve this problem. He felt that the joints could be secured by welding two-inch-thick steel plates over 200 of the joints. Of course, this solution would not be cheap, but it was essential to ensure the integrity of the building.

To resolve this problem, Citicorp would have to be informed. As a first step, LeMessurier consulted with lawyers for his insurance company and for the project's architect. It was decided that LeMessurier and Stubbins would meet with executives from Citicorp to inform them of the problem. They began by meeting with the executive vice president of Citicorp. The chair of Citicorp, Walter Wriston, was then informed. Fortunately, Wriston was very supportive of LeMessurier and decided that Citicorp had to work together with the engineer to ensure that the building would be safe. Two Citicorp vice presidents were assigned the task of managing the repairs.

Plans were immediately drawn up to begin the work, and a fabricator was hired to do the job. A firm was also hired to fit the building with gauges to measure the strain on the individual structural members, and meteorologists were hired to provide daily weather forecasting of expected winds. Next came the delicate task of informing the city building inspectors of what was going on; they would have to approve any plans to alter the structure of the building. The city readily agreed that the changes were required and proceeded to approve the plan. Simultaneously, meetings were held with local disaster-relief agencies to plan for evacuations should a strong storm approach the city.

By now, the newspapers were beginning to hear rumors about the building. But, Citicorp and the City of New York were able to keep the information given to the press to a minimum, which meant that there would be no mass panic about the safety of the building. Fortunately for everyone involved, just as the press began to hear about the problems, a newspaper strike was called, shutting down all of the newspapers in the city. The strike lasted until well after the repairs had been completed.

The welders were able to start the repairs immediately and worked at night to prevent disturbance of the tenants. Work progressed seven days a week and was directed by LeMessurier, who had calculated which joints were most critical and planned the work so that the most essential welds were completed first. All told, the job took about two months. When completed, it was estimated that the building could withstand a storm that was expected only every 700 years and was arguably the most structurally sound building in the city. The total costs for the repairs were never revealed, but they exceeded $8 million; the original cost of the building was $175 million. Surprisingly, Citicorp didn't begin litigation until after the repairs had been made, suing LeMessurier and Stubbins to recover the repair costs. They settled for the $2 million that was the limit of LeMessurier's malpractice insurance.

This case illustrates the benefits of cooperating with persons who came forward after making a mistake: It encourages others to come forward when mistakes have been made and cooperatively work toward a solution. This case also illustrates that rather than losing your reputation when a mistake is made, it can actually be enhanced if you act ethically. LeMessurier sums up this case and the duties of the engineer very beautifully: "You have a social obligation. In return for getting a license and being regarded with respect, you're supposed to be self-sacrificing and look beyond the interests of yourself and your client to society as a whole. And the most wonderful part of my story is that when I did it, nothing bad happened" [Morgenstern, 1995].

The Sealed Beam Headlight Case

Today, nobody worries about the quality of headlamps on automobiles. There are millions of automobiles on the road equipped with headlamps that meet federal safety standards and provide excellent nighttime visibility for the driver. However, this was not always the case. In the early days of the automobile, headlamps were often an unreliable and barely useful part of the vehicle. How unreliable they were became obvious in the early 1930s. By 1933, there were already 24 million motor vehicles operating on the highways in the United States, with over 31,000 fatalities and over 1 million injuries reported [Goodell, 1935]. In 1920, 35% of fatalities occurred during nighttime driving, but this number had risen to 56% by 1933 [Vey, 1935].

In 1935, Paul Goodell, a street-lighting engineer working for the General Illumination Engineering Company, wrote that "visibility has become the weak link in traffic safety, and, as illuminating engineers, we must assume at least a portion of the responsibility in the improvement of traffic hazards. . ." [Goodell, 1935]. Here, we see an engineer urging other engineers to take responsibility for improving the safety of cars, much as modern codes of ethics hold that safety is a paramount concern of the engineer. In fact, the Illumination Engineering Society (IES) in many ways led the way in developing and testing new designs and in working with state and federal regulators to set appropriate standards.

A headlamp consists of three main parts: the light source, a reflector (or reflectors), and a lens. These basic components have remained the same since the invention of automotive lighting through today. Early lamps were housed in a metal box, originally designed to prevent the lamps from being extinguished (they were oil or acetylene flames!), that was later used to protect electric bulbs from damage. Early reflectors were made of highly polished, silvered brass formed into a parabolic shape. Early lenses were made of pressed glass and were used to direct the light in the appropriate direction.

Two main problems existed in these early light designs. First, the silver on the reflector tarnished very easily, leading to diminished headlight intensity. The silver could be polished, but this was difficult to do and was rarely performed by the owner. Low headlight output wasn't only a problem in older cars that had been on the road for many years. A study by Goodell showed that light output was reduced by 60% in automobiles only six months old [Meese, 1982].

The second problem was with the light bulb. The filament had to be located at the focus of the optical system with a very narrow tolerance or the light output would be diminished or misdirected. The variations in the bulbs produced before 1934 made this task very difficult. The problem was mitigated somewhat by the introduction of "prefocused" bulbs in 1934 [Meese, 1982]. But, even when the system was operating correctly, the available brightness of the bulbs was inadequate to permit an automobile to be operated at highway speeds. By the mid-1930s, despite decades of effort, nearly all of the potential performance had been gotten out of the traditional lighting system with still an inadequate lighting situation on the roads [Meese, 1982].

Of course, there were many potential solutions to this problem that were being considered. Fixed lighting of highways was considered. This was a very expensive alternative, involving large up-front capital costs to install lighting along the thousands of miles of highway in the nation. This solution would involve high operating costs for electricity and maintaining the bulbs. It is much less expensive to mount a light on the automobile to deliver illumination on demand, rather than to light a highway all night whether there are cars present or not. Cities could justify such lighting where there is a relatively high traffic density, but highways outside the city were (and still are!) another matter. Other options included severely limiting the amount of driving that could be permitted at night, reducing nighttime speed limits to below 30 mph, or imposing large fines for improper maintenance of automotive lighting systems by the owner. Any new, innovative design for headlamps was sure to be hard for the automobile manufacturers to introduce because during the depression, the high costs of retooling would be very hard to recover.

In 1937, Val Roper, a research engineer at General Electric Company's Automotive Lighting Laboratory in Cleveland, spoke at a meeting of the IES. In his talk, he outlined the requirements for an improved lighting system: a higher wattage bulb; at least two beams, one for open road and the other for use when meeting another car to reduce glare; and the key point, a noticeable difference between the two beams to aid the driver in selecting the correct beam for the driving situation [Meese, 1982]. Roper could make these recommendations in part because he had been working on developing a brighter bulb already.

The reason why brighter bulbs could not be produced was that the filaments could not be sealed adequately. Bright bulbs, in which there was considerable heat generated, developed cracks due to high thermal expansion of the glass. Cracks were especially a problem where the electrical leads entered the glass envelope. The only way to prevent cracking and the resulting bulb failure was to limit the bulb's light output, which reduced the amount of heat generated. In 1935, Roper was working with another lamp inventor at GE, Daniel K. Wright, who had developed a means for placing seals at the point where the electrical leads passed through the glass for use in motion-picture projector bulbs. His design also used borosilicate glass, which was harder than the glass previously used and had a lower coefficient of thermal expansion. These innovations reduced bulb cracking and seemed perfect for application to automotive lamps.

Still, there was a need for improvement in the parabolic reflector. The GE research team reasoned that glass could be used for the shape of the reflector and then could be coated with metal to make it reflective. The whole assembly would then be sealed from the outside environment, thus reducing the problems with tarnishing of the reflector. The problem with this idea was that the technology didn't exist to make a glass surface to the parabolic shape reliably. The GE engineers consulted with Corning Glass Works about this problem. Corning was able to produce a parabolic, aluminized reflector that was more accurate than the conventional design. With the design of an appropriate lens to add to the front surface, the team had developed a far superior lamp [Meese, 1982].

Additional development of this design leading up to 1937 indicated that mass production of this type of headlight was technically feasible, but would be very difficult. It is important to recognize the economic context of this situation. Although there was a huge potential market for such a lamp, there would be substantial extra costs involved. It was not obvious whether the production of this lamp would be financially feasible given the economic situation in the country at the time—the Depression was in full swing.

There was also a potential problem with GE's customers, the headlamp manufacturers. Up until then, GE had supplied the bulbs to these manufacturers for incorporation into their headlamps. This new technology made the headlight a single unit, which might have put these customers out of business. At this point, GE set up a demonstration of its new headlamp for its customers, as well as for the chair of the Engineering Relations Committee of the Society of Automotive Engineers (SAE) and representatives of Ford and General Motors. Of course, this demonstration wasn't necessary, but seemed like the ethical choice, considering the revolutionary nature of the technology. It is interesting to note the names of the automotive light manufacturers present at this demonstration: Guide Lamp, C.M. Hall Company, and Corcoran Brown Company. None of these companies exist today, an indication of how revolutionary the new technology was. As a result of this meeting, the Automobile Manufacturer's Association set up a steering committee to establish standards for headlighting. In this context, it is interesting to note that GE was very generous in its treatment of its customers and others in the use of its sealed beam patents. In fact, GE allowed several manufacturers to consult with their engineers on the design.

While production was being geared up, work began on resolving questions of standardization and regulation of the new design. By 1939, the new standards had been adopted, and the work of the engineers was to help educate state and federal lawmakers who were charged with developing new regulatory standards. The new headlights were introduced in the fall of 1939, and improvements in automotive lighting and highway safety were realized almost immediately.

What are the ethical dimensions of this case? GE could have kept this new technology strictly proprietary. But, realizing the potential for protecting the public safety, the engineers worked with GE management to make the technology as widely available as possible to all lighting and automobile manufacturers. They also worked with regulators and those who developed engineering standards to ensure that this technology would be both accepted engineering practice and required by regulation as soon as possible.

The sealed beam lamp underwent some limited improvement and change during the 40 years after its introduction. However, a new type of design has since been developed in which a high-intensity, replaceable, sealed bulb is a separate component from the reflector and lens of the headlight assembly. The technology to build these bulbs and to easily replace them while protecting the reflector from tarnish has been developed.

Like the Citicorp case, this is an example of engineers doing the right and ethical thing up front and avoiding safety problems and other issues that would later occur. Some innovations that improve safety ought to be shared widely in an industry, even when it means loss of a competitive advantage. This case illustrates what can be done when there is cooperation between industries, professional societies, and the government in trying to solve a problem.

REFERENCES

The Citicorp Center

JOE MORGENSTERN, "The Fifty-nine Story Crisis," *The New Yorker Magazine*, May 29 1995, p. 45.

Sealed Beam Headlights

PAUL H. GOODELL, "Street Lighting and the Science of Seeing," *Transactions of the I.E.S.*, Jan. 1935, p. 50.

GEORGE P. E. MEESE, "The Sealed Beam Case: Engineering in the Public and Private Interest," *Business and Professional Ethics Journal*, Spring 1982, vol. 1, pp. 1–23.

ARNOLD H. VEY, "Relation of Highway Lighting to Highway Accidents," *Transactions of the I.E.S.*, Jan. 1935, p. 83.

Problems

The Citicorp Center

1. The Citicorp Center met the applicable standards and city codes. What might have gone wrong in the design process for this building?

2. What went right in the aftermath of the discovery of the problem?

3. What might have happened if Citicorp had immediately sued?

4. What role did the newspaper strike have in the successful outcome of this case? Would things have been different had there been more press scrutiny?

5. Is it acceptable to try to keep newspapers in the dark about this type of problem?

6. Should there have been full disclosure of the hazards of the building to people who worked in the building and people in surrounding neighborhoods? Is the answer the same even if emergency-response agencies were well informed and an evacuation plan was in place?

7. Use line drawing to examine the possible alternatives that LeMessurier had when he discovered that the building was not as strong as it should have been. Identify other alternatives that he had, and decide if there were other ethically acceptable possibilities.

Sealed Beam Headlight

8. Did GE have to inform its customers of the new technology? Did it have to inform the SAE of it?

9. What obligation did GE have to try to overcome the difficulties with regulations and standards? Did it have an economic interest in seeing these standards adopted?

10. Could this type of industry-wide solution to a public safety problem occur in today's economic and legal environment?

11. On August 21, 1998, newspapers began to report that General Motors was about to introduce an infrared sensor and display system in its Cadillac models, starting in the 2000 model

year. This is a heat-sensitive system that displays images of people or animals in the darkness in front of an automobile. The display is a small screen on the windshield. The range is up to 500 yards; regular headlights have a range of about 100 yards. Research the developments in this technology and see if GM approached the introduction of this technology in the same way that GE approached the introduction of the sealed beam headlight. Has GM taken the most ethical approach?

Appendix A Codes of Ethics of Professional Engineering Societies

THE INSTITUTE OF ELECTRICAL AND ELECTRONICS ENGINEERS, INC.*

We, the members of the IEEE, in recognition of the importance of our technologies affecting the quality of life throughout the world, and in accepting a personal obligation to our profession, its members and the communities we serve, do hereby commit ourselves to the highest ethical and professional conduct and agree:

1. to accept responsibility in making engineering decisions consistent with the safety, health and welfare of the public, and to disclose promptly factors that might endanger the publicor the environment;
2. to avoid real or perceived conflicts of interest whenever possible, and to disclose them to affected parties when they do exist;
3. to be honest and realistic in stating claims or estimates based on available data;
4. to reject bribery in all its forms;
5. to improve the understanding of technology, its appropriate application, and potential consequences;
6. to maintain and improve our technical competence and to undertake technological tasks for others only if qualified by training or experience, or after full disclosure of pertinent limitations;
7. to seek, accept, and offer honest criticism of technical work, to acknowledge and correct errors, and to credit properly the contributions of others;
8. to treat fairly all persons regardless of such factors as race, religion, gender, disability, age, or national origin;
9. to avoid injuring others, their property, reputation, or employment by false or malicious action;
10. to assist colleagues and co-workers in their professional development and to support them in following this code of ethics.

Approved by the IEEE Board of Directors, August 1990

NSPE CODE OF ETHICS FOR ENGINEERS

Preamble

Engineering is an important and learned profession. As members of this profession, engineers are expected to exhibit the highest standards of honesty and integrity. Engineering has a direct and vital impact on the quality of life for all people. Accordingly, the services provided by engineers require honesty, impartiality, fairness and equity, and must be dedicated to the protection of the public health, safety, and welfare. Engineers must perform under a standard of professional behavior that requires adherence to the highest principles of ethical conduct.

I. Fundamental Canons

Engineers, in the fulfillment of their professional duties, shall:

1. Hold paramount the safety, health and welfare of the public.
2. Perform services only in areas of their competence.
3. Issue public statements only in an objective and truthful manner.
4. Act for each employer or client as faithful agents or trustees.
5. Avoid deceptive acts.
6. Conduct themselves honorably, responsibly, ethically, and lawfully so as to enhance the honor, reputation, and usefulness of the profession.

II. Rules of Practice

1. Engineers shall hold paramount the safety, health, and welfare of the public.
    a. If engineers' judgment is overruled under circumstances that endanger life or property, they shall notify their employer or client and such other authority as may be appropriate.
    b. Engineers shall approve only those engineering documents that are in conformity with applicable standards.
    c. Engineers shall not reveal facts, data or information without the prior consent of the client or employer except as authorized or required by law or this Code.
    d. Engineers shall not permit the use of their name or associate in business ventures with any person or firm that they believe are engaged in fraudulent or dishonest enterprise.
    e. Engineers having knowledge of any alleged violation of this Code shall report thereon to appropriate professional bodies and, when relevant, also to public authorities, and cooperate with the proper authorities in furnishing such information or assistance as may be required.
2. Engineers shall perform services only in the areas of their competence.
    a. Engineers shall undertake assignments only when qualified by education or experience in the specific technical fields involved.
    b. Engineers shall not affix their signatures to any plans or documents dealing with subject matter in which they lack competence, nor to any plan or document not prepared under their direction and control.

c. Engineers may accept assignments and assume responsibility for coordination of an entire project and sign and seal the engineering documents for the entire project, provided that each technical segment is signed and sealed only by the qualified engineers who prepared the segment.

3. Engineers shall issue public statements only in an objective and truthful manner.
   a. Engineers shall be objective and truthful in professional reports, statements, or testimony. They shall include all relevant and pertinent information in such reports, statements, or testimony, which should bear the date indicating when it was current.
   b. Engineers may express publicly technical opinions that are founded upon knowledge of the facts and competence in the subject matter.
   c. Engineers shall issue no statements, criticisms, or arguments on technical matters that are inspired or paid for by interested parties, unless they have prefaced their comments by explicitly identifying the interested parties on whose behalf they are speaking, and by revealing the existence of any interest the engineers may have in the matters.
4. Engineers shall act for each employer or client as faithful agents or trustees.
   a. Engineers shall disclose all known or potential conflicts of interest that could influence or appear to influence their judgment or the quality of their services.
   b. Engineers shall not accept compensation, financial or otherwise, from more than one party for services on the same project, or for services pertaining to the same project, unless the circumstances are fully disclosed and agreed to by all interested parties.
   c. Engineers shall not solicit or accept financial or other valuable consideration, directly or indirectly, from outside agents in connection with the work for which they are responsible.
   d. Engineers in public service as members, advisors, or employees of a governmental or quasi-governmental body or department shall not participate in decisions with respect to services solicited or provided by them or their organizations in private or public engineering practice.
   e. Engineers shall not solicit or accept a contract from a governmental body on which a principal or officer of their organization serves as a member.
5. Engineers shall avoid deceptive acts.
   a. Engineers shall not falsify their qualifications or permit misrepresentation of their or their associates' qualifications. They shall not misrepresent or exaggerate their responsibility in or for the subject matter of prior assignments. Brochures or other presentations incident to the solicitation of employment shall not misrepresent pertinent facts concerning employers, employees, associates, joint venturers, or past accomplishments.
   b. Engineers shall not offer, give, solicit or receive, either directly or indirectly, any contribution to influence the award of a contract by public authority, or which may be reasonably construed by the public as having the effect of intent to influencing the awarding of a contract. They shall not offer any gift or other valuable consideration in order to secure work. They shall not pay a commission, percentage, or brokerage fee in order to

secure work, except to a bona fide employee or bona fide established commercial or marketing agencies retained by them.

III. Professional Obligations

1. Engineers shall be guided in all their relations by the highest standards of honesty and integrity.
   a. Engineers shall acknowledge their errors and shall not distort or alter the facts.
   b. Engineers shall advise their clients or employers when they believe a project will not be successful.
   c. Engineers shall not accept outside employment to the detriment of their regular work or interest. Before accepting any outside engineering employment they will notify their employers.
   d. Engineers shall not attempt to attract an engineer from another employer by false or misleading pretenses.
   e. Engineers shall not actively participate in strikes, picket lines, or other collective coercive action.
   f. Engineers shall not promote their own interest at the expense of the dignity and integrity of the profession.
2. Engineers shall at all times strive to serve the public interest.
   a. Engineers shall seek opportunities to participate in civic affairs; career guidance for youths; and work for the advancement of the safety, health and well-being of their community.
   b. Engineers shall not complete, sign, or seal plans and/or specifications that are not in conformity with applicable engineering standards. If the client or employer insists on such unprofessional conduct, they shall notify the proper authorities and withdraw from further service on the project.
   c. Engineers shall endeavor to extend public knowledge and appreciation of engineering and its achievements.
3. Engineers shall avoid all conduct or practice that deceives the public.
   a. Engineers shall avoid the use of statements containing a material misrepresentation of fact or omitting a material fact.
   b. Consistent with the foregoing, Engineers may advertise for recruitment of personnel.
   c. Consistent with the foregoing, Engineers may prepare articles for the lay or technical press, but such articles shall not imply credit to the author for work performed by others.
4. Engineers shall not disclose, without consent, confidential information concerning the business affairs or technical processes of any present or former client or employer, or public body on which they serve.
   a. Engineers shall not, without the consent of all interested parties, promote or arrange for new employment or practice in connection with a specific project for which the Engineer has gained particular and specialized knowledge.
   b. Engineers shall not, without the consent of all interested parties, participate in or represent an adversary interest in connection with a specific

project or proceeding in which the Engineer has gained particular specialized knowledge on behalf of a former client or employer.

5. Engineers shall not be influenced in their professional duties by conflicting interests.
   a. Engineers shall not accept financial or other considerations, including free engineering designs, from material or equipment suppliers for specifying their product.
   b. Engineers shall not accept commissions or allowances, directly or indirectly, from contractors or other parties dealing with clients or employers of the Engineer in connection with work for which the Engineer is responsible.
6. Engineers shall not attempt to obtain employment or advancement or professional engagements by untruthfully criticizing other engineers, or by other improper or questionable methods.
   a. Engineers shall not request, propose, or accept a commission on a contingent basis under circumstances in which their judgment may be compromised.
   b. Engineers in salaried positions shall accept part-time engineering work only to the extent consistent with policies of the employer and in accordance with ethical considerations.
   c. Engineers shall not, without consent, use equipment, supplies, laboratory, or office facilities of an employer to carry on outside private practice.
7. Engineers shall not attempt to injure, maliciously or falsely, directly or indirectly, the professional reputation, prospects, practice, or employment of other engineers. Engineers who believe others are guilty of unethical or illegal practice shall present such information to the proper authority for action.
   a. Engineers in private practice shall not review the work of another engineer for the same client, except with the knowledge of such engineer, or unless the connection of such engineer with the work has been terminated.
   b. Engineers in governmental, industrial, or educational employ are entitled to review and evaluate the work of other engineers when so required by their employment duties.
   c. Engineers in sales or industrial employ are entitled to make engineering comparisons of represented products with products of other suppliers.
8. Engineers shall accept personal responsibility for their professional activities, provided, however, that Engineers may seek indemnification for services arising out of their practice for other than gross negligence, where the Engineer's interests cannot otherwise be protected.
   a. Engineers shall conform with state registration laws in the practice of engineering.
   b. Engineers shall not use association with a nonengineer, a corporation, or partnership as a "cloak" for unethical acts.
9. Engineers shall give credit for engineering work to those to whom credit is due, and will recognize the proprietary interests of others.
   a. Engineers shall, whenever possible, name the person or persons who may be individually responsible for designs, inventions, writings, or other accomplishments.

b. Engineers using designs supplied by a client recognize that the designs remain the property of the client and may not be duplicated by the Engineer for others without express permission.
c. Engineers, before undertaking work for others in connection with which the Engineer may make improvements, plans, designs, inventions, or other records that may justify copyrights or patents, should enter into a positive agreement regarding ownership.
d. Engineers' designs, data, records, and notes referring exclusively to an employer's work are the employer's property. Employer should indemnify the Engineer for use of the information for any purpose other than the original purpose.

As Revised July 1996

"By order of the United States District Court for the District of Columbia, former Section 11(c) of the NSPE Code of Ethics prohibiting competitive bidding, and all policy statements, opinions, rulings or other guidelines interpreting its scope, have been rescinded as unlawfully interfering with the legal right of engineers, protected under the antitrust laws, to provide price information to prospective clients; accordingly, nothing contained in the NSPE Code of Ethics, policy statements, opinions, rulings or other guidelines prohibits the submission of price quotations or competitive bids for engineering services at any time or in any amount."

Statement by NSPE Executive Committee

In order to correct misunderstandings which have been indicated in some instances since the issuance of the Supreme Court decision and the entry of the Final Judgment, it is noted that in its decision of April 25, 1978, the Supreme Court of the United States declared: "The Sherman Act does not require competitive bidding."

It is further noted that as made clear in the Supreme Court decision:

1. Engineers and firms may individually refuse to bid for engineering services.
2. Clients are not required to seek bids for engineering services.
3. Federal, state, and local laws governing procedures to procure engineering services are not affected, and remain in full force and effect.
4. State societies and local chapters are free to actively and aggressively seek legislation for professional selection and negotiation procedures by public agencies.
5. State registration board rules of professional conduct, including rules prohibiting competitive bidding for engineering services, are not affected and remain in full force and effect. State registration boards with authority to adopt rules of professional conduct may adopt rules governing procedures to obtain engineering services.
6. As noted by the Supreme Court, "nothing in the judgment prevents NSPE and its members from attempting to influence governmental action . . . "

NOTE: In regard to the question of application of the Code to corporations vis-à-vis real persons, business form or type should not negate nor influence conformance of individuals to the Code. The Code deals with professional services, which services must be performed by real persons. Real persons in turn establish and implement policies within business structures. The Code is clearly written to apply to the Engineer and items incumbent on members of NSPE to endeavor to live up to its provisions. This applies to all pertinent sections of the Code.

NOTE: There is also the NSPE Ethics Reference Guide, which fleshes out some of this stuff.

AMERICAN SOCIETY OF MECHANICAL ENGINEERS

Ethics

ASME requires ethical practice by each of its members and has adopted the following Code of Ethics of Engineers as referenced in the ASME Constitution, Article C2.1.1.

Code of Ethics of Engineers

The Fundamental Principles

Engineers uphold and advance the integrity, honor and dignity of the engineering profession by:

I. Using their knowledge and skill for the enhancement of human welfare;

II. Being honest and impartial, and serving with fidelity the public, their employers and clients; and

III. Striving to increase the competence and prestige of the engineering profession.

1. Engineers shall hold paramount the safety, health and welfare of the public in the performance of their professional duties.
2. Engineers shall perform services only in the areas of their competence.
3. Engineers shall continue their professional development throughout their careers and shall provide opportunities for the professional and ethical development of those engineers under their supervision.
4. Engineers shall act in professional matters for each employer or client as faithful agents or trustees, and shall avoid conflicts of interest or the appearance of conflicts of interest.
5. Engineers shall build their professional reputation on the merit of their services and shall not compete unfairly with others.
6. Engineers shall associate only with reputable persons or organizations.
7. Engineers shall issue public statements only in an objective and truthful manner.

The ASME Criteria for Interpretation of the Canons

The ASME criteria for interpretation of the Canons are guidelines and represent the objectives toward which members of the engineering profession should strive. They are principles which an engineer can reference in specific situations. In addition, they provide interpretive guidance to the ASME Board on Professional Practice and Ethics on the Code of Ethics of Engineers.

1. Engineers shall hold paramount the safety, health and welfare of the public in the performance of their professional duties.
   a. Engineers shall recognize that the lives, safety, health and welfare of the general public are dependent upon engineering judgments, decisions and practices incorporated into structures, machines, products, processes and devices.

b. Engineers shall not approve or seal plans and/or specifications that are not of a design safe to the public health and welfare and in conformity with accepted engineering standards.

c. Whenever the Engineers' professional judgments are over ruled under circumstances where the safety, health, and welfare of the public are endangered, the Engineers shall inform their clients and/or employers of the possible consequences.

(1) Engineers shall endeavor to provide data such as published standards, test codes, and quality control procedures that will enable the users to understand safe use during life expectancy associated with the designs, products, or systems for which they are responsible.

(2) Engineers shall conduct reviews of the safety and reliability of the designs, products, or systems for which they are responsible before giving their approval to the plans for the design.

(3) Whenever Engineers observe conditions, directly related to their employment, which they believe will endanger public safety or health, they shall inform the proper authority of the situation.

d. If engineers have knowledge of or reason to believe that another person or firm may be in violation of any of the provisions of these Canons, they shall present such information to the proper authority in writing and shall cooperate with the proper authority in furnishing such further information or assistance as may be required.

2. Engineers shall perform services only in areas of their competence.

a. Engineers shall undertake to perform engineering assignments only when qualified by education and/or experience in the specific technical field of engineering involved.

b. Engineers may accept an assignment requiring education and/or experience outside of their own fields of competence, but their services shall be restricted to other phases of the project in which they are qualified. All other phases of such project shall be performed by qualified associates, consultants, or employees.

3. Engineers shall continue their professional development throughout their careers, and should provide opportunities for the professional and ethical development of those engineers under their supervision.

4. Engineers shall act in professional matters for each employer or client as faithful agents or trustees, and shall avoid conflicts of interest or the appearance of conflicts of interest.

a. Engineers shall avoid all known conflicts of interest with their employers or clients and shall promptly inform their employers or clients of any business association, interests, or circumstances which could influence their judgment or the quality of their services.

b. Engineers shall not undertake any assignments which would knowingly create a potential conflict of interest between themselves and their clients or their employers.

c. Engineers shall not accept compensation, financial or otherwise, from more than one party for services on the same project, or for services pertaining to the same project, unless the circumstances are fully disclosed to, and agreed to, by all interested parties.

d. Engineers shall not solicit or accept financial or other valuable considerations, for specifying products or material or equipment suppliers, without disclosure to their clients or employers.

e. Engineers shall not solicit or accept gratuities, directly or indirectly, from contractors, their agents, or other parties dealing with their clients or employers in connection with work for which they are responsible. Where official public policy or employers' policies tolerate acceptance of modest gratuities or gifts, engineers shall avoid a conflict of interest by complying with appropriate policies and shall avoid the appearance of a conflict of interest.

f. When in public service as members, advisors, or employees of a governmental body or department, Engineers shall not participate in considerations or actions with respect to services provided by them or their organization(s) in private or product engineering practice.

g. Engineers shall not solicit an engineering contract from a governmental body or other entity on which a principal, officer, or employee of their organization serves as a member without disclosing their relationship and removing themselves from any activity of the body which concerns their organization.

h. Engineers working on codes, standards or governmental sanctioned rules and specifications shall exercise careful judgment in their determinations to ensure a balanced viewpoint, and avoid a conflict of interest.

i. When, as a result of their studies, Engineers believe a project(s) will not be successful, they shall so advise their employer or client.

j. Engineers shall treat information coming to them in the course of their assignments as confidential, and shall not use such information as a means of making personal profit if such action is adverse to the interests of their clients, their employers or the public.

   (1) They will not disclose confidential information concerning the business affairs or technical processes of any present or former employer or client or bidder under evaluation, without his consent, unless required by law or court order.

   (2) They shall not reveal confidential information or finding of any commission or board of which they are members unless required by law or court order.

   (3) Designs supplied to Engineers by clients shall not be duplicated by the Engineers for others without the express permission of the client(s).

k. Engineers shall act with fairness and justice to all parties when administering a construction (or other) contract.

l. Before undertaking work for others in which Engineers may make improvements, plans, designs, inventions, or other records which may justify seeking copyrights, patents, or proprietary rights, Engineers shall enter into positive agreements regarding the rights of respective parties.

m. Engineers shall admit their own errors when proven wrong and refrain from distorting or altering the facts to justify their mistakes or decisions.

n. Engineers shall not accept professional employment or assignments outside of their regular work without the knowledge of their employers.

o. Engineers shall not attempt to attract an employee from other employers or from the market place by false or misleading representations.

5. Engineers shall build their professional reputation on the merit of their services and shall not compete unfairly with others.
   a. Engineers shall negotiate contracts for professional services on the basis of demonstrated competence and qualifications for the type of professional service required.
   b. Engineers shall not request, propose, or accept professional commissions on a contingent basis if, under the circumstances, their professional judgments may be compromised.
   c. Engineers shall not falsify or permit misrepresentation of their, or their associates, academic or professional qualification. They shall not misrepresent or exaggerate their degrees of responsibility in or for the subject matter of prior assignments. Brochures or other presentations used to solicit personal employment shall not misrepresent pertinent facts concerning employers, employees, associates, joint venturers, or their accomplishments.
   d. Engineers shall prepare articles for the lay or technical press which are only factual. Technical Communications for publication (theses, articles, papers, reports,etc.) which are based on research involving more than one individual (including students and supervising faculty, industrial supervisor/researcher or other co-workers) must recognize all significant contributors. Plagiarism, the act of substantially using another's ideas or written materials without due credit, is unethical. (See Appendix.)
   e. Engineers shall not maliciously or falsely, directly or indirectly, injure the professional reputation, prospects, practice or employment of another engineer, nor shall they indiscriminately criticize another's work.
   f. Engineers shall not use equipment, supplies, laboratory or office facilities of their employers to carry on outside private practice without consent.
6. Engineers shall associate only with reputable persons or organizations.
   a. Engineers shall not knowingly associate with or permit the use of their names or firm names in business ventures by any person or firm which they know, or have reason to believe, are engaging in business or professional practices of a fraudulent or dishonest nature.
   b. Engineers shall not use association with non-engineers, corporations, or partnerships to disguise unethical acts.
7. Engineers shall issue public statements only in an objective and truthful manner.
   a. Engineers shall endeavor to extend public knowledge, and to prevent misunderstandings of the achievements of engineering.
   b. Engineers shall be completely objective and truthful in all professional reports, statements or testimony. They shall include all relevant and pertinent information in such reports, statements or testimony.
   c. Engineers, when serving as expert or technical witnesses before any court, commission, or other tribunal, shall express an engineering opinion only when it is founded on their adequate knowledge of the facts in issue, their background of technical competence in the subject matter, and their belief in the accuracy and propriety of their testimony.

d. Engineers shall issue no statements, criticisms, or arguments on engineering matters which are inspired or paid for by an interested party, or parties, unless they preface their comments by identifying themselves, by disclosing the identities of the party or parties on whose behalf they are speaking, and by revealing the existence of any financial interest they may have in matters under discussion.

e. Engineers shall be truthful in explaining their work and merit, and shall avoid any act tending to promote their own interest at the expense of the integrity and honor of the profession or another individual.

8. Engineers accepting membership in The American Society of Mechanical Engineers by this action agree to abide by this Society Policy on Ethics and procedures for its implementation.

Responsibility: Council on Member Affairs/Board on Professional Practice and Ethics

Adopted: March 7, 1976
Revised: December 9, 1976
December 7, 1979
November 19, 1982
June 15, 1984
(editorial changes 7/84)
June 16, 1988
September 12, 1991
September 11, 1994

AMERICAN SOCIETY OF CIVIL ENGINEERS*

The Engineering Code of Ethics

Fundamental Principles

Engineers uphold and advance the integrity, honor and dignity of the engineering profession by:

using their knowledge and skill for the enhancement of human welfare;
being honest and impartial and serving with fidelity the public, their employers and clients;
striving to increase the competence and prestige of the engineering profession; and
supporting the professional and technical societies of their disciplines.

Fundamental Canons

Engineers shall hold paramount the safety, health and welfare of the public in the performance of their professional duties.
Engineers shall perform services only in areas of their competence.
Engineers shall issue public statements only in a subjective and truthful manner.
Engineers shall act in professional matters for each employer or client as faithful agents or trustees, and shall avoid conflicts of interest.
Engineers shall build their professioal reputation on the merit of their services and shall not compete unfairly with others.

*Courtesy of ASCE

Engineers shall act in such a manner as to uphold and enhance the honor, integrity, and dignity of the engineering profession.

Engineers shall continue their professional development throughout their careers, and shall provide opportunities for the professional development of those engineers under their supervision.

AMERICAN INSTITUTE OF CHEMICAL ENGINEERS

AIChE Code of Ethics

American Institute of Chemical Engineers

The Council of the American Institute of Chemical Engineers adopted this Code of Ethics to which it expects that the professional conduct of its members shall conform, and to which every applicant attests by signing his or her membership application.

Members of the American Institute of Chemical Engineers shall uphold and advance the integrity, honor, and dignity of the engineering profession by: being honest and impartial and serving with fidelity their employers, their clients, and the public; striving to increase the competence and prestige of the engineering profession; and using their knowledge and skill for the enhancement of human welfare. To achieve these goals, members shall:

1. Hold paramount the safety, health, and welfare of the public in performance of their professional duties.
2. Formally advise their employers or clients (and consider further disclosure, if warranted) if they perceive that a consequence of their duties will adversely affect the present or future health or safety of their colleagues or the public.
3. Accept responsibility for their actions and recognize the contributions of others; seek critical review of their work and offer objective criticism of the work of others.
4. Issue statements or present information only in an objective and truthful manner.
5. Act in professional matters for each employer or client as faithful agents or trustees, and avoid conflicts of interest.
6. Treat fairly all colleagues and co-workers, recognizing their unique contributions and capabilities.
7. Perform professional services only in areas of their competence.
8. Build their professional reputations on the merits of their services.
9. Continue their professional development throughout their careers, and provide opportunities for the professional development of those under their supervision.

Appendix B Bibliography

GENERAL BOOKS ON ENGINEERING ETHICS

Charles E. Harris, Jr., Michael S. Pritchard, and Michael J. Rabins, *Engineering Ethics, Concepts and Cases*, Wadsworth Publishing Company, Belmont, CA, 1995.

Deborah G. Johnson, *Ethical Issues In Engineering*, Prentice-Hall, Englewood Cliffs, NJ, 1991.

Mike W. Martin and Roland Schinzinger, *Ethics in Engineering*, 2d. ed., McGraw-Hill, New York, 1989.

JOURNALS WITH ARTICLES ON ENGINEERING ETHICS AND CASES

Business and Professional Ethics Journal, published by the Center for Applied Philosophy, University of Florida, Gainesville, FL.

Journal of Business Ethics, Kluwer Academic Publishers.

WEB SITES

There are many Web sites that deal with issues of engineering ethics. Some of the better ones are listed here. Of course, the addresses of Web sites tend to be very unstable, so these addresses may no longer be valid at the time you read this text. Many of these Web sites offer links to other Web-based ethics resources as well.

Ethics Center for Engineering and Science, MIT: *http://web.mit.edu/ethics/www/*

Society on Social Implications of Technology (SSIT) of the IEEE: *http://www4.ncsu.edu/unity/users/j/jherkert/index.html*

Applied Ethics Resources, Univ. of Calgary: *http://www.ucalgary.ca/UofC/faculties/HUM/Philosophy/appliedethics.html*

Department of Science and Technology Studies, Cornell University: *http://www.sts.cornell.edu/Lilly2.html*

Index